Interindividual
Behavioral Variability
in Social Insects

Interindividual Behavioral Variability in Social Insects

EDITED BY

Robert L. Jeanne

Routledge
Taylor & Francis Group

LONDON AND NEW YORK

First published 1988 by Westview Press, Inc.

Published 2018 by Routledge
52 Vanderbilt Avenue, New York, NY 10017
2 Park Square, Milton Park, Abingdon, Oxon OX14 4RN

Routledge is an imprint of the Taylor & Francis Group, an informa business

Library of Congress Cataloging-in-Publication Data
Interindividual behavioral variability in social
   insects.
   (Westview studies in insect biology)
   Includes index.
   1. Insects--Behavior. 2. Insect societies.
I. Jeanne, Robert L. II. Series.
QL496.I422   1988   595.7'051   86-13278

ISBN 13: 978-0-367-01061-4 (hbk)
ISBN 13: 978-0-367-16048-7 (pbk)

Contents

Preface

A key element of social organization in the
insects is polyethism, or division of labor among
specialized castes within the colony. Polyethism
manifests itself not only as division of labor
between the reproductives and the workers, but in
most species as specialization among worker
subcastes as well. Worker specialization is often
age-based, a worker changing roles as it ages. In
some ant and most termite species there is also
morphological specialization of workers for
different roles. The many excellent empirical and
theoretical studies addressing caste have given us
a good understanding of caste differentiation,
behavior, ergonomics, and evolution.
     At the level of the whole colony or caste,
the behavioral response to a contingency such as
nest overheating or an attack by a predator is
virtually deterministic. But while it is possible
to predict how the colony will respond in a given
situation, predicting how an individual in the
colony will respond in the same situation is quite
another matter. Not only is the expression of
behavior at the individual level probabilistic,
but individual behavior profiles and activity
levels often vary dramatically, independently of
caste and age. It is not farfetched to assert
that no two colony members behave alike during
their lifetimes.
     In recent years increasing attention has been
given to the questions raised by these features of
social insect behavior at the individual level.
That the members of a caste are not all

behaviorally identical automatons is not surprising, but why does interindividual variability, often extreme, seem to be the rule? Is the variability really just noise in the system, or is it somehow adaptive at the colony level? And given the seemingly idiosyncratic behavior of the individuals making up the colony, how can behavior of the colony itself be anything but chaotic?

In 1985 the Entomological Society of America and the North American Section of the International Union for the Study of Social Insects cosponsored a symposium to address the issue of individual variability in the social insects. The symposium was convened in December at the annual ESA national meeting in Hollywood, Florida. The chapters in this volume (with the exception of chapter 9) evolved out of papers presented at that symposium.

Most of the papers included herein represent empirical studies of some aspect of the phenomenon of variability in social insect behavior. They illustrate the range of ways colony members can differ from one another and interpret the variability in terms of the external environment, social context, or individual experience. Others take some preliminary steps toward modelling the phenomenon. It will be seen that we are still a long way from having answers to many of the questions addressed--and posed--in this volume, but a map of the territory in which the answers lie is coming into focus, and it is challenging and interesting terrain indeed. If the book succeeds in kindling in the reader an interest in the role of individual variability in the behavior of social insects, then its purpose will have been achieved.

It is a pleasure to thank the Entomological Society of America and the North American Section of the International Union for the Study of Social Insects for contributing funds toward the realization of the symposium. Without their help, many of the participants would not have been able to attend. I also thank those who helped in preparing this book for publication. Gregg Henderson helped with proofreading. For typing I thank the Department of Entomology at the

University of Wisconsin and especially Sue Eder, who did an expert job on the final drafts. The staff of Westview Press always cheerily answered my questions about editorial details.

Robert L. Jeanne

1

Individual Traits of
Social Behavior in Ants

*Pierre Jaisson, Dominique Fresneau,
and Jean-Paul Lachaud*

INTRODUCTION

A century after Charles Darwin, researchers
are still tackling important problems raised by
insect societies. One example is the localization
of the focus of selective pressure. An important
consequence of the division of reproductive roles
in social insects is that both individual and
group selection are involved in their evolution.
The reproducers (males and fertile females) are
directly affected in their reproductive function
by individual selection. But the reproducers are
also indirectly affected by the adaptive value of
the neuters in their interactions with the
environment (this is one of the forms that group
selection takes in social insects). The inclusive
fitness of a given society is determined by its
reproductive success, which is in turn a result of
the impact of these selection pressures on the
group members.
Apart from certain rare budding species, ant
queens are alone when the society is founded.
This is, therefore, a particularly important
moment. Selection, forcibly acting directly on
the individual at this stage, could alter the
future of the whole society. This is probably
why, during the evolution of the ants, we can see
a progressive change from non-claustral to totally
claustral foundation. The queens thus avoid this
potential source of selection by not leaving the
nest to search for food for themselves or for the
first brood generation. In many species belonging

1

to the primitive ant-subfamilies Myrmeciinae and
Ponerinae, the founding queen forages outside the
nest (see review in Wilson 1971). The partially-
claustral formula is scarce but not totally absent
in more evolved genera, occurring in the
myrmicines Acromyrmex octospinosa Reich (Cordero
1963), Manica rubida Latr. (Le Masne and Bonavita
1969), Strumigenys rufobrunea Santschi and
Smithistruma emarginata (Mayr) (Dejean 1982), and
in the formicine Cataglyphis cursor nigra Andre
(Fridman and Avital 1983).

Claustral foundation, which is the general
rule in these evolved subfamilies, results in an
increasing separation between castes. The queen
survives by the histolysis of the alary muscles,
which, following swarming, are redundant. This
food reserve is sufficient to support the founding
female for several months. When the society is
formed predators often exercise a strong selective
pressure on foragers when they leave the nest.
Field studies of marked individuals of the
formicine Cataglyphis bicolor Fabricius, carried
out by Schmid-Hempel and Schmid-Hempel (1984) have
shown that the life expectancy is only 6.1 days
after the first foray out of the nest. In the
laboratory the same foragers can have an average
life-span of several months (Schmid-Hempel 1983;
Wehner et al. 1983).

The vulnerability of a forager to predation
results in the society's losing part of its food
supply. The interest of the individual is thus
completely linked to that of the society. This
fact makes a joint holistic and individual
approach necessary when studying ant societies, as
in the case of other social insects.

THE STRUCTURE OF AN ANT SOCIETY

Apart from the males, whose only function is
to produce females, the society is composed of two
female castes: the queens and the workers (Figure
1.1). Workers may be subdivided into different
physical subcastes, into which individuals fall in
a discrete or continuous fashion. However, in
some species there is no distinguishable
difference between queens and workers. For

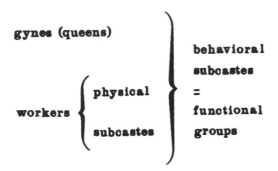

Figure 1.1. Female castes and subcastes in social Hymenoptera.

example, in the ponerine Ophthalmopone berthoudi Forel, studied by Peeters (1982), all individuals are ergatomorph and can be mated. The workers, and in certain cases the queens, are also distributed into behavioral subcastes, with given social roles in the society.

The taxonomic level of a given ant species does not seem to be strictly correlated with the degree of the queen's integration into the behavioral subcastes. This has been emphasized by Fresneau et al. (1982). They compared the participation of the queen in social tasks in three species belonging to the presumably primitive subfamily Ponerinae (Neoponera obscuricornis Emery and Ectatomma tuberculatum Olivier) and to the evolved subfamily Myrmicinae (Myrmecina graminicola Fabricius). The Neoponera queens are involved in the division of labor of the society, but in the other ponerine, Ectatomma, the queen is totally separate from the worker force and she does not belong to the behavioral subcastes. These data suggest that the ponerine group is heterogeneous from a behavioral point of view, and includes both primitive and evolved species. An explanation of the behavioral separation of female castes might be the queen numbers. However, in the evolved and monogynous Myrmecina species studied by these authors, there is a clear participation by the queen in the behavioral subcastes. She is mainly involved in the caring activities (principally oriented towards the eggs). A comparison of the dimorphism between queen and worker in these three species showed that the queen/worker size ratio seems to be related to the degree of queen participation in the social tasks (see Figure 1.2). Paradoxically, M. graminicola is in an intermediate position, behaviorally and morphologically, between the well integrated queens of N. obscuricornis and the non-integrated queens of E. tuberculatum. The authors concluded that the morphological separation of the castes would be a better explanation of the non-integration of the queen in the division of labor than the species' membership in a so-called "evolved" or "primitive" subfamily.

METHODS AND APPROACHES TO THE STUDY OF INDIVIDUAL TRAITS OF SOCIAL BEHAVIOR IN ANTS

Relatively few methods are available. One method consists of isolating an ant in order to examine its behavioral potential in comparison with other individuals outside of the regulatory

| | QUEEN / WORKER | | Nursing activities of the queen |
	HEAD Interocular width (mm)	THORAX Prothoracic width (mm)	
(PONERINAE) *NEOPONERA OBSCURICORNIS*	1.052	1.219	.640 (60 obs.)
ECTATOMMA TUBERCULATUM	1.531	1.917	.007 (70 obs.)
MYRMECINA GRAMINICOLA	1.165	1.521	.370 (90 obs.)
(MYRMICINAE)			

Figure 1.2. Relationship between queen/worker dimorphism and the nursing activities performed by the queen in two ponerine and one myrmicine species. (Modified from Fresneau et al. 1982.)

social pressure of the colony. When the
conditions are strictly controlled this method
enables us to approach the different intrinsic
individual predispositions (idiosyncrasy) in
isolated conditions shown by members of the same
colony. However, one must be cautious because the
isolation of an essentially social insect could
alter its behavior.

Less open to criticism, but more difficult to
carry out, is the study of the individual ant in
the context of the colony. This approach needs a
marking method. When the species has a high
number of individuals per colony a genuine
individual study is impossible. However, it is
possible to follow a group of individuals sharing
a characteristic: this is the group marking
method. In such cases, group "marking" may simply
consist of morphological traits which can be
correlated with age. Examples of this natural
group marking based on physical characters are:

1. Pigmentation: In many ant species the
   callow is clear and it may be hours, days,
   or weeks before the mature pigmentation
   appears. For example in various Myrmica
   species there are several weeks between the
   light yellow post-eclosion stage and the
   dark reddish brown terminal coloration (see
   Weir 1958).
2. Membership in a physical subcaste: In
   polymorphic species soldiers can be studied
   in comparison with minor workers.
3. Degree of wear of the mandibles: Well-
   shaped at birth, the teeth are
   progressively worn with age. This natural
   group marking has been used for example by
   Smeeton (1982) in Myrmica (see Figure 1.3).
4. Degree of salissure of the cuticle: This
   natural marking related with age has
   recently been used by Wilson and Hölldobler
   (1986) in the genus Basiceros, where the
   increasing salissure gives the older
   individuals an effective camouflage.

One problem with natural group marking is the
difficulty of discriminating two adjacent classes
when the physical character used is continuously

6

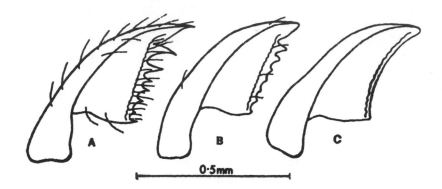

Figure 1.3. Mandible wear in <u>Myrmica</u> <u>rubra</u>
    workers (right mandibles shown). (A) Newly
    ecloded; mandible teeth sharp. (B) Old;
    mandible teeth worn. (C) Old; mandible teeth
    worn down to the base. (From Smeeton 1982).

distributed in the colony (for example,
pigmentation). This drawback is absent when
artificial marking of a homogeneous group of
individuals is used. This technique may involve
application of a permanent paint to adults or
larvae, the clipping of claws, tarsi, or
exoskeletal spines, or the ingestion of colored
foods.
     When the species includes a reduced number of
individuals per colony or when foundation is being
studied, a genuine individual study becomes
feasible. This approach inevitably involves
artificial marking, and can necessitate a large
number of different marks. These marks can
consist of blobs of paint differing in number,
color, or position; or copper wires tied around
the body, which may vary in color, position, or
even the form of the knot (e.g., Provost 1983).
Other methods include coded mutilations (number of
tarsi clipped, which leg clipped, etc... Lenoir
1979a; Morel 1983). The use of coded labels is
increasing. They can be made of nickel (Verron
and Barreau 1974) or of photographic film
(Fresneau and Charpin 1977) stuck to the ant's
thorax or abdomen. This coded label carries a

number, letter, or symbol (see Figure 1.4).

When a class of individuals or an individual ant is identified in its society, how can its role be analyzed? In the framework of an investigation of the individual and its relationships with the society, two levels of analysis have been used. First, the purely <u>descriptive</u> level, which results in a catalogue of behavioral acts performed by each species, caste, subcaste and individual. It consists of identifying and recording the frequency of each behavior and characterizing the individual or the group of individuals responsible for it. This can lead to the production of ethograms (or sociograms if marking is done on a genuinely individual basis). Using this method, the frequency distribution of different acts amongst individuals of different sizes and ages can also be studied. Descriptive studies have been carried out with two different recording methods, which are linked to the marking method:

1. The <u>sporadic sampling</u> method, where the behavior is recorded at random. This method is often combined with the method of statistical analysis suggested by Fagen and Goldman (1977). This calculation can result in an estimation of the "real" number of kinds of behavioral elements (or acts) in a given species from the total number of acts observed and the total number of behavioral categories. This estimation of the "real" number of behavioral elements depends on the grid of behavioral acts arbitrarily chosen by the investigator and generally excludes inactivity as an important trait of the individual behavioral profile (see p. 9).
2. The <u>systematic scanning</u> method, where each behavior performed by each individual is recorded. This method is more useful in colonies with a reduced number of individuals or for the study of identified individuals in the context of their society. The systematic scanning method lends itself especially to the methods of numerical taxonomy, notably the algorithms of hierarchical classification. Individual

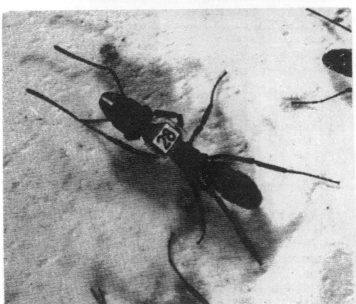

Figure 1.4. Individually marked workers with photographed labels. (A) <u>Neoponera</u> <u>apicalis</u>; (B) <u>Ectatomma</u> <u>ruidum</u>.

inactivity, which appears to play an important role, can easily be recorded by this method.

Whatever particular method is used, it must be able to: (1) identify the behavioral acts involved, (2) record the frequency of each behavior, (3) record each contribution to the frequency of the behavior.

After the descriptive stage, the analysis can be continued at an experimental level, where one uses results from the descriptive level of study in order to investigate the origins of behavior. In particular, this method allows us to study the ethogenesis of the individual ant, the mechanisms of social regulation, and the physiological correlates of the division of social roles.

THE BEHAVIORAL REPERTOIRES PERFORMED BY INDIVIDUALS

The results obtained by the "sporadic sampling method" and those obtained by the "systematic scanning method" may sometimes produce the same kind of information. This is the case for species-specific behavioral repertoires and ethograms, since they are based on an overall approach to the society's structure, for example, by taking into account the presence or absence and frequency of certain behavioral patterns.

Fagen and Goldman's (1977) method has led various authors to conclude that the behavioral repertoires they obtain are genuinely representative. However, a survey of the literature shows that there are major problems with this approach that have not been sufficiently studied. Nevertheless, these problems appear clearly when one tries to compare some of the ethograms obtained by the same author or by different authors for different species or even for the same species.

In the genus Pheidole, for example, the compilation of the results described for six species (minor and major data grouped) gives a grid of 51 behavioral acts from which 17 appear to be species-specific, since they are present for

one species only (Table 1.1). The same can be
observed for Leptothorax (Table 1.2) where, in a
grid of 50 behavioral acts, 23 are species-
specific. Moreover, in this last genus, the
problem is more complex since for a given species
the repertoire can vary with the colony studied:
in three L. ambiguus Emery colonies, 14 behavioral
acts (from a total of 49) were recorded in only
one of the colonies. A similar variation was
observed in L. longispinosus Roger where 6 acts
(from a total of 28) were seen in only one colony
out of the four studied. Given that these
experiments were conducted by the same author it
seems that there is an important intercolony
behavioral variation.

When observations are made by different
authors this variability is even greater.
Zacryptocerus varians Fr. Smith is a typical
example of this (Table 1.3). A compilation of the
results obtained by Wilson (1976a) and by Cole
(1980) produces a global repertoire of 47
behavioral acts, of which 24 are not common to the
two studies. Both authors obtained similar values
for the total behavioral repertoire using the
Fagen and Goldman test, thus suggesting that the
repertoires are similar (Cole 1980). However, the
data actually obtained do not lead to such a
conclusion. Further, the behavioral grid
resulting from the data compilation is clearly
more important than the theoretical values. Such
problems raise doubts about the validity of
species-specific repertoires as defined by the
Fagen and Goldman method.

As emphasized by Wilson (1976b), these
repertoires are, within certain limits, arbitrary
since behavioral patterns may be regrouped or
subdivided. Thus, the repertoire size, which
depends on the criteria of the observer, will be
increased or decreased accordingly. With this
kind of data analysis it is always difficult to
interpret the absence of a particular behavior.
It may result from a genuine absence of the
behavior, or it may be that the author did not
take it into account in the experimental situation
chosen for the observations. This problem is
highlighted by those studies in which behavior
patterns which are normally infrequently observed

Table 1.1

Comparison of behavioral repertoires (minor and major) within some species of the genus Pheidole

Behavioral acts	P.p.	P.j.	P.m.	P.e.	P.d.	P.h.	Sum
Self groom	X	X	X	X	X	X	6
Trophallaxis w/minor	X	X	X	X	X	X	6
Forage	X	X	X	X	X	X	6
Lick/groom larva	X	X	X	X	X	X	6
Eat solid food inside nest	X	X	X	X	X	X	6
Carry larva(e)	X	X	X	X	X	X	6
Trophallaxis w/larva	X	X	X	X	X	X	6
Allogroom minor	X	X	X	X	X	X	6
Carry egg(s)	X	X	X	X	X	X	6
Carry pupa(e)	X	X	X	X	X	X	6
Lick/groom egg	X	X	X	X	X	X	6
Lick/groom pupa	X	X	X	X	X	X	6
Allogroom major	X	X	X	X	X	X	6
Assist pupal eclosion	X	X	X	X	X	X	6
Eat dead adult	X		X	X	X	X	5
Carry food to/inside nest	X	X	X	X	X		5
Feed outside nest	X	X	X	X	X		5
Assist larval eclosion	X	X		X	X	X	5
Trophallaxis w/major	X		X	X	X	X	5
Carry dead adult	X		X	X	X	X	5
Feed larva solid food	X	X	X	X	X		5

(continued)

Table 1.1 (continued)

Behavioral acts	P.p.	P.g.	P.m.	P.e.	P.d.	P.h.	Sum
Allogroom queen	X	X	X	X		X	5
Trophallaxis w/queen	X	X	X	X		X	5
Handle nest material/refuse	X	X	X	X	X		5
Chew on seed	X	X	X	X			4
Carry adult nestmate	X	X	X		X		4
Antennal tipping	X	X	X	X			4
Carry dead larva/pupa	X	X	X	X			4
Lay odor trail	X	X	X	X			4
Handle meconium	X	X	X				3
Lick wall nest	X		X	X			3
Feed on larva/pupa	X	X	X				3
Guard		X		X		X	3
Eject infrabuccal pellet and dispose		X		X			2
Lick meconium					X		1
Allogroom w/alate female	X						1
Allogroom w/male		X					1
Allogroom w/alate female/male	X						1
Extrude anal region for licking				X			1
Anal trophallaxis	X						1
Regurgitate w/alate female	X						1

	37	31	31	33	25	28	51
Regurgitate w/alate female/ male					X		1
Aggression	X						
Carry brood exuvia						X	1
Carry meconium						X	1
Carry nest material						X	1
Eat brood exuvia						X	1
Guard food outside nest				X			1
Patrol at food						X	1
Patrol at arena						X	1
Retrieve food						X	1
Total behavioral acts observed	37	31	31	33	25	28	51

P. pubiventris (P.p.), P. guilelmimuelleri (P.g.) and P. megacephala (P.m.) from Wilson 1984; P. embolopyx (P.e.) from Wilson and Hölldobler 1985; P. hortensis (P.h.) from Calabi et al. dentata Mayr. (P.d.) from Wilson 1976b; P. 1983.

Table 1.2
Comparison of behavioral repertoires within different colonies of two species of the genus <u>Leptothorax</u>

Behavioral acts	L.a1	L.a2	L.a3	L.11	L.12	L.14	L.13	Sum L.a	Sum L.1	Sum
Selfgroom	X	X	X	X	X	X	X	3	4	7
Groom larva	X	X	X	X	X	X	X	3	4	7
Regurgitate w/larva	X	X	X	X	X	X	X	3	4	7
Groom pupa	X	X	X	X	X	X	X	3	4	7
Regurgitate w/worker	X	X	X	X	X	X	X	3	4	7
Lick nest wall	X	X	X	X	X	X	X	3	4	7
Forage	X	X	X	X	X	X	X	3	4	7
Rest	X	X	X	X	X	X	X	3	4	7
Inspect larva	X	X	X	X	X	X	X	3	4	7
Antennate w/worker	X	X	X	X	X	X	X	3	4	7
Antennate body	X	X	X	X	X	X	X	3	4	7
Feed insect inside nest	X	X	X	X	X	X	X	3	4	7
Inspect pupa	X	X	X	X	X	X	X	3	4	7
Carry larva	X	X	X	X		X	X	3	3	6
Allogroom worker	X	X	X	X	X		X	3	3	6
Handle nest material	X	X	X		X	X	X	3	3	6

Behavior									
Carry pupa	X	X	X	X	X	X	3	3	6
Inspect egg	X	X	X	X	X	X	3	3	6
Antennate queen	X	X	X	X	X	X	2	4	6
Carry egg	X	X	X	X			3	2	5
Groom egg	X	X	X	X			3	1	4
Be groomed by worker	X	X		X	X	X	2	2	4
Inspect nest entrance	X	X	X		X		3	1	4
Drink	X	X	X	X	X		3	1	4
Regurgitate w/queen	X	X		X			2	1	3
Fight other workers			X	X	X	X	0	3	3
Feed outside nest	X	X	X				3	0	3
Lay egg	X	X	X				3	0	3
Move inside nest	X	X	X				3	0	3
Carry exuvium	X	X	X				3	0	3
Eat exuvium	X	X	X				3	0	3
Move outside	X	X	X				3	0	3
Inspect prey	X	X	X				3	0	3
Carry prey	X	X	X				3	0	3
Feed larva solid		X		X	X		1	1	2
Allogroom queen		X		X	X		1	1	2
Assist larval ecdysis	X	X	X				2	0	2
Inspect exuvium	X	X	X				2	0	2
Assist adult eclosion	X	X					1	0	1

(continued)

Table 1.2 (concluded)

Behavioral acts	L.a1	L.a2	L.a3	L.11	L.12	L.14	L.13	Sum L.a	Sum L.1	Sum
Be carried/carry live nestmate			X					1	0	1
Fight queen	X							1	0	1
Regurgitate w/male	X							1	0	1
Assist egg laying		X						1	0	1
Antennate male	X							1	0	1
Allogroom male	X							1	0	1
Carry male	X							1	0	1
Allogroom w/alate female			X					1	0	1
Carry refuse	X							1	0	1
Eat prey	X							1	0	1
Inspect food	X							1	0	1
Total behavioral acts observed	43	38	33	22	21	21	20			
Total behavioral specific acts	49	49	49	28	28	28	28	49	28	63

L. a1, a2, a3 = 3 L. ambiguus societies (from Herbers 1983); L. 11, 12, 13, 14 = 4 L. longispinosus societies (from Herbers 1982).

Table 1.3
Comparison of behavioral repertoires within the
species Zacryptocerus varians

Behavioral Acts	Cole (1980)	Wilson (1976a)	Sum
Self grooming	X	X	2
Antennal tipping	X	X	2
Lick wall of nest	X	X	2
Allogroom minor	X	X	2
Allogroom major	X	X	2
Allogroom queen	X	X	2
Lay trophic egg	X	X	2
Carry or manipulate egg(s)	X	X	2
Lick egg(s)	X	X	2
Carry or manipulate larva(e)	X	X	2
Lick larva(e)	X	X	2
Assist larval ecdysis	X	X	2
Carry or manipulate pupa(e)	X	X	2
Lick pupa(e)	X	X	2
Feed egg to larva	X	X	2
Feed infrabuccal pellet to larva	X	X	2
Receive/solicit abdominal trophallaxis from minor	X	X	2
Receive/solicit abdominal trophallaxis from major	X	X	2
Receive/solicit abdominal trophallaxis from queen	X	X	2
Regurgitate w/larva	X	X	2
Regurgitate w/minor	X	X	2
Regurgitate w/major	X	X	2
Regurgitate w/queen	X	X	2
Lay normal egg		X	1
Carry dead insect		X	1
Carry infrabuccal pellet	X		1
Extrude infrabuccal pellet		X	1
Carry solid remains	X		1
Carry dead nestmate		X	1
Assist ecdysis to pupa		X	1
Assist eclosion to adult		X	1
Feed on solid food inside nest		X	1
Feed on insect remains	X		1

(continued)

Table 1.3 (concluded)

Behavioral Acts	Cole (1980)	Wilson (1976a)	Sum
Feed on larva	X		1
Feed on egg	X		1
Feed on infrabuccal pellet	X		1
Feed on solid remains	X		1
Share infrabuccal pellet	X		1
Forage outside nest		X	1
Feed on honey outside nest		X	1
Feed on insect outside nest		X	1
Extrude sting and/or anal tube		X	1
Lay odor trail to food		X	1
Jittering		X	1
Feed trophic egg to minor		X	1
Cannibalism on larva		X	1
Excavating		X	1
Total theoretical behavioral acts	40	42	–
Total behavioral acts observed	31	39	47

(From Wilson 1976a and Cole 1980).

are presented in great detail. This may lead to
their being given undue importance. For example,
the inclusion of alates has resulted in the
discovery of five "new" behavioral categories in
Leptothorax ambiguus (Herbers 1983; see Table 1.2)
and four in Pheidole pubiventris Mayr (Wilson
1984; see Table 1.1).
    Given the difficulties, we must be cautious
about comparative studies based on species-
specific behavioral repertoires. Unfortunately no
common behavioral grid exists which allows us to
compare data interspecifically. Indeed, the
possibility of such a grid being developed seems
unlikely.

INDIVIDUAL BEHAVIOR AND SOCIAL ORGANIZATION

The study of individual behavioral repertoires does not produce a satisfactory comparative characterization of the species. Perhaps a better method is to study how the behavior of the individual is integrated into the organization of the society.

The distribution of individuals in the social group in relation to their behavioral repertoire has been described by taking into account the physical subcaste, the age class, or the behavioral subcaste to which they belong.

Individual Behavior, Polymorphism, and Social Organization

In species with physical subcastes, soldiers are the simplest to study because their morphology results in a straightforward natural marking. Some species show a clear dimorphic worker force with a minor and a major (or soldier) form. This is the case in the genus Pheidole, studied by Wilson (1984). Using the sporadic sampling method, Wilson showed that in ten species of this genus the fraction of majors in the colony varied from 3 to 25 percent. The behavioral acts performed by these soldiers did not show a great diversity. The repertoire size of the minor caste was generally larger than 30 (as estimated by the Fagen and Goldman method). However, the author found that the number of different behavioral elements performed by the soldiers tended to increase with an increase in the proportion of soldiers in the colony, the repertoire size comprising between 4 and 19 acts (see Figure 1.5). This result is consistent with Wilson's theory that the higher the numerical representation of a physical subcaste, the greater the exploitation of the investment which it represents for the colony (Wilson 1968; Oster and Wilson 1978). This hypothesis, developed particularly for anatomically specialized individuals, is important in understanding the ergonomics of physical and behavioral subcastes.

Wilson (1985a) recently estimated that 83

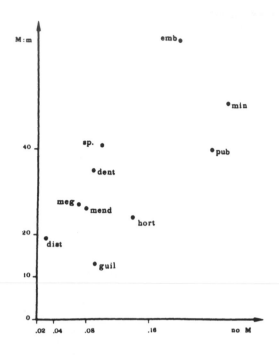

Figure 1.5.   The relation between the fraction of
     major workers in colonies (no. M) and the
     ratio of the number of behavioral acts by
     major workers to the number of behavioral
     acts by minor workers (M:m).   Filled circles
     refer to species of <u>Pheidole</u> (dent = <u>dentata</u>;
     dist = <u>distorta</u>; emb = <u>embolopyx</u>; guil =
     <u>guilelmimuelleri</u>; hort = <u>hortensis</u>; meg =
     <u>megacephala</u>; mend = <u>mendicula</u>; min =
     <u>ninutula</u>; pub = <u>pubiventris</u>; sp. = undeter-
     mined.   (Modified from Wilson 1984.)

percent of the 363 living ant genera listed by
Brown (1973) are entirely monomorphic (that is,
none of the component species has a major
subcaste). Most of the polymorphic species have
monogynous colonies. We believe that there is an
evolutionary relationship between polymorphism and
monogyny (Figure 1.6). The development of
monogyny can be understood in terms of the
advantage gained in concentrating investment in
certain successful genomes. However, monogyny has
the disadvantage of reducing interindividual
genetic variability within the society, which is

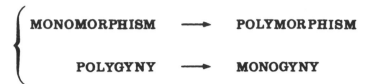

Figure 1.6. Possible correlation between the
    evolution of worker polymorphism and the
    evolution of monogyny in ants.

important in the division of labor and in the
potential evolutionary flexibility of the
species. This could be sufficient to explain the
evolution of a mechanism providing a new source of
interindividual variability in some of the
monogynous genera: worker polymorphism. A recent
study of differences in worker size between
monogynous and polygynous colonies in the fire ant
Solenopsis invicta Buren (Greenberg et al. 1985)
supports this interpretation. In contrast to
Greenberg et al.'s expectation, it appears that
polygynous colonies of this species are less
polymorphic than monogynous nests. Moreover,
major workers are produced only by monogynous
colonies, in spite of a lower genetic
interindividual variability (Ross and Fletcher
1985).

Individual Behavior, Age, and Social Organization

        Where there is no caste polymorphism amongst
the workers (this is the case in most species),
the age of the workers is the essential element
determining the division of roles in the society.
        Following the studies of Otto (1958), Weir
(1958), and Dobrzanska (1959), it is clear that as
the workers grow older they change roles, usually
progressing from nurse to forager. The effects of
age polyethism are additional to those produced by
the system of morphological castes in highly
polymorphic species, such as Pheidole. This led
Wilson (1976b) to study a "Temporal Caste System"
that was superimposed on the "Physical Caste
System." The "Temporal Caste" was defined as an

age-group of individuals performing a particular task for a sustained period of time. As in the case where there is a discontinuity of morphological castes, Wilson believes that the evolution of the ants has produced two organizational alternatives for the "Temporal Caste System": the continuous and discontinuous systems. In a continuous system the society has to multiply its number of behavioral subcastes as the number of tasks to be carried out increases. This problem does not arise in the discontinuous system, where each age group has its own particular social task. However, the continuous or discontinuous nature of age polyethism is based both on the nomenclature of the behavioral elements considered as defining a given task, and on an estimation of the time scale during which the workers become specialized in a given task.

The fact that individuals may shift from one task to another fundamentally differentiates age polyethism from the statistically rigid polyethism linked to physical caste. This shows the complexity of the problem no matter what recording method is adopted. In order to overcome some of these difficulties, Wilson (1976b) proposed a three-stage methodology. First, the elaboration of an ethogram, second, the determination of the age of the workers based on their colour, and third, a measure of polyethism based on a range of ages previously selected. Calabi et al. (1983) used this method for their study of Pheidole hortensis Forel, which provides an excellent example of this approach. Having carefully defined five shades of pigmentation for both majors and minors, the authors observed a whole colony for ten weeks. On the basis of these observations they produced a genuine polyethic map using 20 behavioral criteria. By calculating an index of specialization (RPM), they were able to compare the performance of each age group (see Figure 1.7).

If we consider that category I represents the youngest level of pigmentation and category V the oldest, it is clear that there is an age-linked polyethism in the two physical subcastes of P. hortensis. In the minors, stage I workers limit their activity to behavioral acts 1-6, stage II to

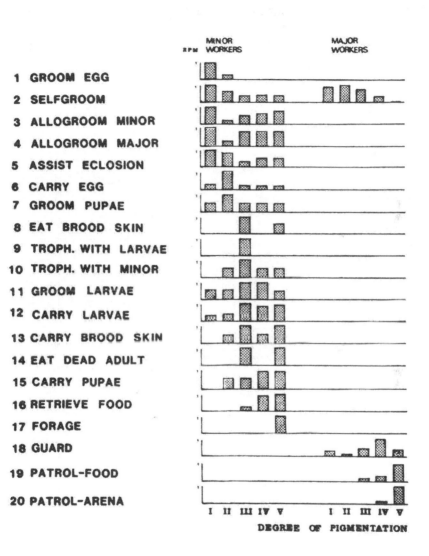

Figure 1.7. Relative performance measures (RPM)
of various behavioral acts in the repertoires
of major and minor workers of Pheidole
hortensis. Roman numerals correspond to the
degree of pigmentation and thus approximate
to the age classes. (Modified from Calabi et
al. 1983.)

behavioral acts 8-14 and stage V to behavioral
acts 12-17. The majors perform only four acts,
and only show a specialization for guard duty at
stage IV and for foraging at stage V. The authors
state that the older stages maintain the level of
performance of behavioral acts which characterized
their younger stages. Thus, the foragers of stage
V continue their nursing and allogrooming
activities. It is as though the workers enrich
their behavioral repertoire without ever
completely losing previously-present behavior.
This "atypical" schema was explained as being the
result of a fixation of the caring tasks, due to
the functional needs of the colony, which had
recently substantially increased its production of
offspring. But it is perhaps more likely to have
been due to a behavioral variability in older
individuals. Because the authors did not study
individual behavioral acts they were unable to
discount this hypothesis.

This example clearly shows the theoretical
and practical limits of the global analysis of age
polyethism. It can provide us with information
about the effect of age on the "average"
individual, but it does not reveal the variety of
behavioral patterns shown by individuals in the
same age class. Furthermore, estimation of the age
of individuals, based only on pigmentation, cannot
be applied to all species. In many monomorphic
species, adult pigmentation is achieved in the
first few days after eclosion; thus it is very
probable that in many cases the "oldest" stage of
pigmentation will, in fact, be comprised of
workers of several different ages.

Individual Behavior, Behavioral Subcastes, and Social Organization

The sytematic scanning method used is based
on an analysis of the behavior of individual ants
aimed at revealing the collective functioning of
the society.

Lenoir (1979a) used this method to study the
division of labor in young Lasius niger Linné
colonies. The data were analysed using
correspondence analysis (Benzecri 1973). This

method produces a simplified description of the
associations between ants and behaviors. However,
while this procedure is useful it is extremely
complex and unwieldy. It was further developed by
Fresneau (1984), Lachaud and Fresneau (1985), and
Perez-Bautista et al. (1986) for primitive
species. In the following example, taken from a
study of the social organization in an adult
colony of <u>Ectatomma ruidum</u> Roger, a Neotropical
ponerine (Corbara et al. 1986), there are three
principal steps:

1. Individual marking of all members of the
    society with numbers stuck on the thorax
    (Fresneau and Charpin 1977). This
    preparatory period lasted one month in
    order for the precise age of the
    individuals eclosing during this opening
    stage to be known. The colony consisted of
    one queen, one dealate female, six males
    and 72 workers. All types of brood were
    present--20 eggs, 60 larvae, and 45 pupae.
2. For three days photos of the whole society
    were taken by an automatic camera. On the
    basis of these photographs a behavioral
    frequency profile was constructed for each
    individual. In this example, each ant was
    observed 101 times, making a total of more
    than 8000 observations for the colony as a
    whole.
3. A sociogram for the society was constructed
    by grouping twelve behavioral categories.
    The data were then analysed using a
    numerical taxonomic method based on an
    hierarchical classification algorithm. The
    results were summarized in a dendrogram
    (Figure 1.8), which defines the behavioral
    subcastes of the society by classifying
    individual ants by the proximity of their
    behavioral profiles. Six groups could thus
    be characterized. Amongst them, the queen
    was clearly distinct (group one, on the
    left).

In order to produce a social ethogram
outlining the behavioral characteristics of the
different behavioral subcastes, the authors

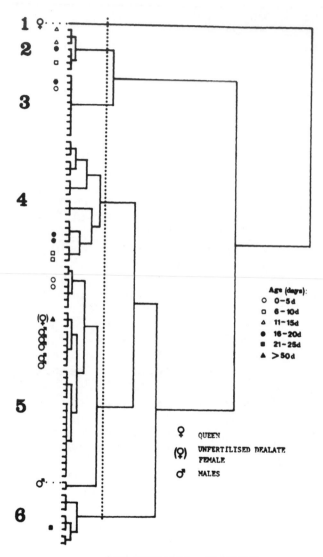

ECTATOMMA   RUIDUM

Figure 1.8.   Dendrogram showing the classification
of 79 ants found in an adult colony of
Ectatomma ruidum.   (Modified from Corbara et
al. 1986.)

applied Bertin's (1977) method (see Figure 1.9).
The six groups of the dendrogram were grouped in
columns, the twelve behavioral categories in

rows. The combination of rows and columns enables us to arrive at a composite picture of the social organization of the colony solidly based on the observed data. We can thus see that each group has a behavioral spectrum which does not correspond to the total number of behavioral acts seen in the colony. There is a continuous behavioral evolution from subcaste 1 (the queen) to subcaste 6, from nursing to foraging. Whatever the composition of the brood, nursing seems to be characterized by an emphasis on nursing of larvae as opposed to nursing of eggs and to a lesser extent as opposed to nursing of pupae. This is not generally found in other ponerine species. The strong separation of the queen from other ants in the ethogram seems to be a characteristic of the advanced ponerines such as the species of the genus Ectatomma. In Neoponera (Fresneau 1984; Perez-Bautista et al. 1986) the queens form a part of the behavioral subcastes with the workers.

The intermediate subcaste 4 and group 5 perform higher levels of inactivity and nonsocial behavior (exploration, self-grooming, and feeding) inside the nest. It is interesting to note that inactivity can be an important characteristic of the more densely populated behavioral subcaste of an ant society. The social roles of the behavioral subcastes are deduced from their profiles:

1. Subcaste 1 includes only the queen and is characterized by nursing of the eggs and larvae.
2. Subcaste 2 is characterized by a high level of nursing of larvae (principally small larvae).
3. Subcaste 3 also nurses larvae (large larvae) and pupae.
4. Subcaste 4 nurses pupae and with subcaste 3 initiates allogrooming. It is also occupied with nest-maintenance.
5. Subcaste 5 is essentially characterized by a very passive social role and receives more allogrooming. These ants seem to be oriented towards activities outside the nest. All males as well as the unfertilized female are included in this

28

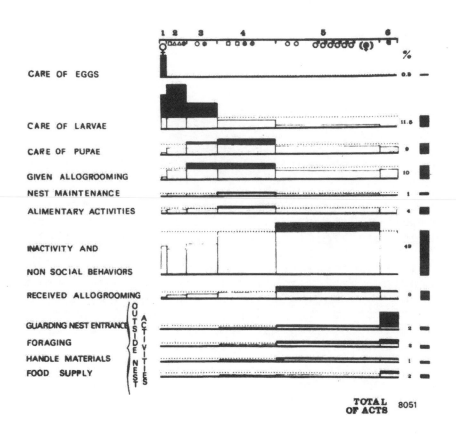

BEHAVIORAL SUBCASTES

Figure 1.9. Social ethogram of behavioral castes (in columns) in an adult colony of Ectatomma ruidum. The ethogram was constructed by assigning to each column a value corresponding to the total number of acts it contained; the level of each behavior was then calculated as a proportion that would be expected for each behavioral category if the behavioral acts were equally distributed amongst groups. Each value greater than this level is marked in black, in order to show the degree of specialization of each group of ants. (Modified from Corbara et al. 1986.)

group.
6. Subcaste 6 guards the nest entrance,
   forages, and supplies food for the colony.

It is surprising to note that of the sample
of 14 ants of known age the 0-5 day old workers
are found in group 5 (socially inactive) and group
3 (nurses of larvae), given that such ants are
normally thought to show the same behavior.
Similarly, 6-10 day old workers can be both nurses
of larvae (group 2) and involved in nursing of
cocoons and nest maintenance (group 4). Finally,
16-20 day old workers are found in groups 2, 3 and
4, that is, they show a behavioral repertoire
stretching from the nursing of larvae to nest
maintenance.
These results, although based on a small
sample, show clearly that the classical schema of
age polyethism has only a statistical value. The
proportion of older individuals (more than 25 days
old, not shown on the dendrogram) is greater in
groups 4 and 5, which have social roles relatively
distant from nursing. But we need to note also
the large heterogeneity of behaviors shown by the
younger age groups. The importance of this
phenomenon needs to be investigated; accurate
control and measurement of the age and
physiological state of each member of the society
would be a first step.

INDIVIDUAL BEHAVIOR AND SOCIAL REGULATION

The ability of insect societies to adapt
themselves to a severe depletion in the size of a
subcaste was named "régulation sociale" by Grassé
(1946) or "social homeostasis" by Emerson
(1956). The individual analysis of the society
provides the best possibility of explaining and,
if possible, predicting the dynamics of the social
regulation mechanisms. Experimental studies on
ants have supplied important results in this
area. These experiments principally consist of
(1) modifying the minor/major subcaste ratio, (2)
reducing the total number of individuals in a
colony, and (3) removing a behavioral subcaste
(and eventually reintroducing the individuals in

the colony after a period of time).

Modification of the Minor/Major Subcastes Ratio

Wilson (1984) removed a large number of minor workers from individual colonies of three different species of Pheidole (guilelmimuelleri Forel, megacephala Fabricius, and pubiventris Mayr). A rapid behavioral homeostatic response was shown by all societies. For example, when the minor/major ratio dropped below 1:1 (the normal ratio ranging from 3:1 to 20:1, depending on the species) the rate of social activity of the major subcaste increased from 1.4 to 4.5.

The same author recently obtained a similar result in the other dimorphic myrmicine genus Erebomyrma (Wilson 1986). Here the repertoire size increased to 3-5 times and the activity level rose about 10 times.

As the author pointed out, in both genera the variation in the size of the repertoire performed by the major subcaste does not exist in the minor subcaste, despite large experimental variations in the minor/major ratio. Another interesting result is that the ergonomic resiliency of the colony (as it is termed by Wilson [1984]) takes place soon after the removal of the minors (about 1 hour). This resiliency is considered to be a consequence of the behavioral "elasticity" of the individuals of the major physical subcaste. However, only the use of the scanning sampling technique could reveal interindividual variation and its role in the behavioral elasticity shown by the major subcaste.

Reduction of the Total Number of Individuals in a Colony

An example of this method was given by Wilson (1983) in the leaf-cutter ant Atta cephalotes Linné. Using sporadic sampling, he demonstrated that the number of different behavioral acts performed by individuals depends on their size (as measured by head-width). Wilson also carried out experimental studies using the reduction of the

size of the colony. In each of four adult
colonies the population was reduced from about
10,000 to 236 workers. Further, a juvenile size-
frequency distribution (identical to a natural
colony less than a year old) was imposed in the
reduced societies. (In Atta cephalotes the worker
size-frequency is continuously distributed.)

The first brood cycle produced by these
artificially reduced colonies showed worker pupae
with a size-frequency distribution similar to that
of a young colony. For example, the frequency of
individuals with a head-width of 1.8 mm before the
reduction was between 8.0 and 43.2 percent,
depending on the colony. After the population
reduction this same size class comprised merely 0-
2.3 percent of the newly produced workers. This
interesting result shows that the physical
characteristics of the produced individuals (and
thus their behavioral potentials) depend on the
size of the colony and not its age. In normal
conditions the four colonies would have produced a
higher proportion of individuals with head-width
1.8 mm. The interpretation presented by Wilson is
that naturally founded or artificially reduced
colonies have the same optimizing strategy: they
produce a maximum number of individuals performing
a wider range of essential tasks and with fewer
interindividual differences.

Removal of a Behavioral Subcaste

A third method of studying the mechanisms of
social regulation is sociotomy. This involves
depriving the colony of one or more of its
behavioral subcastes (for ants of a certain age),
and then observing how the society copes with this
change. This method was used by Lenoir (1979b) in
his study of Tapinoma erraticum Latreille. The
absence of older workers following sociotomy led
to a more rapid development of the younger
workers: they tended to become foragers earlier
than normal. By isolating different behavioral
subcastes in Lasius niger, this author was further
able to show that the behavioral plasticity of
workers declines with age: older ants are largely
specialists in foraging. Lenoir suggests that

social regulation in these species depends upon the flexibility of the younger ants, the nurses, which are able to show an accelerated behavioral development.

The hypothesis has been investigated by McDonald and Topoff (1985) in Novomessor albisetosus Mayr. In undisturbed colonies of this species workers become specialized foragers at about 68 days of age. If all the foragers are removed from the society, this maturation period is reduced to a mere 19 days. The response of the isolated foragers is to revert almost immediately to nursing behaviors. The authors suggest that the presence of the brood plays a central role in the acceleration of individual development. In particular, the presence of newly-eclosed workers seems to induce the shift to foraging behavior in young "domestics".

The systematic study of individual behavior allows us to investigate the mechanisms of social regulation in the light of the role and variation of the individual. A classic approach is to carry out a three-phase study: First the undisturbed period as a control, then sociotomy, during which both the colony and the isolated individuals are observed, and finally the reinsertion of the isolated individuals into the colony. Such a study was carried out on a colony of Neoponera apicalis Latreille by Fresneau and Lachaud (1985). This study lasted 51 days and consisted of three phases:

1. The control phase, P1 (21 days), was used to measure the initial level of activity corresponding to each task.
2. In the experimental phase, P2 (10 days), the eight most active nurses were separated, together with a part of the brood. The 28 other ants were kept with the rest of the brood. During this period a maximum amount of social regulation may be expected.
3. During the second control phase, P3 (20 days), the colony was reunited.

Figure 1.10 summarizes the behavioral evolution of the eight separated nurses and the rest of the colony during the three phases of the experiment. During sociotomy (phase 2) each subcolony behaviorally compensated for the missing individuals. The separated nurses spent some of their time on tasks outside the nest, but returned to full nursing status in phase 3. The other subcolony, deprived of nurses, restablished a level of nursing activity comparable to that observed before separation. However, after the reunification of the colony these ants returned to their original profiles. In all cases the change in behavior towards nursing was made at the expense of "non-social" behaviors. In the subcolony deprived of nurses only seven ants became nurses ("new-nurses") and were thus involved in social regulation. The behavioral evolution of the seven "new-nurses" and the other individuals is shown in Figure 1.11. The behavior of the "new-nurses" during the control phase (P1) shows that at this stage they were part of the inactive group, although they also engaged in nursing. The other ants, however, maintained a stable behavioral profile in all three phases and did not participate in social regulation.

Not all individuals respond to the pressure of social regulation in the same manner. We can generally predict their responses to sociotomy on the basis of their behavior immediately beforehand. The division of labor is characterized by a high degree of inertia. This is shown by the fact that all the ants returned to their original behavioral status following reunification of the colony. This functional inertia is probably based upon the interaction of physiological, ethogenetic, and social factors. It needs to be studied in greater detail.

PHYSIOLOGICAL CORRELATES OF INDIVIDUAL BEHAVIOR

It is possible that intra-individual behavioral variation may be linked to age-related physiological changes. The behavioral evolution of an individual is generally correlated with the progressive maturation of anatomical and physiological features. For example, in the genus Formica young workers, which work inside the nest

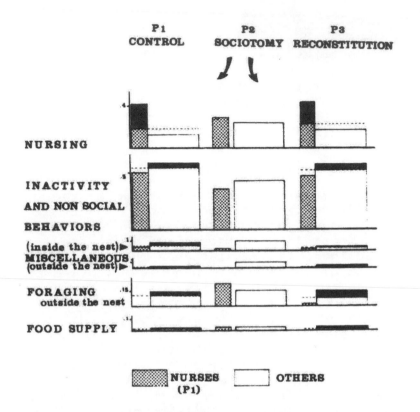

Figure 1.10. Behavioral profiles of nurses and
other ants during the three periods studied
expressed as relative frequencies of total
number of acts in each subgroup (on
ordinate). The dotted lines indicated the
colony's mean level of activity for each
period. By convention, that portion of each
histogram that exceeds this level is shown in
black and indicates the degree of
specialization in that task by the
subgroup. (Modified from Fresneau and
Lachaud 1985.)

and carry out "domestic" or "nursing" functions,
have highly-developed ovaries and fat bodies.
They also have active post-pharyngeal and labial
glands. On the other hand, older workers, which
are generally foragers, show an important
regression of the ovaries (Otto 1958; Dobrzanska

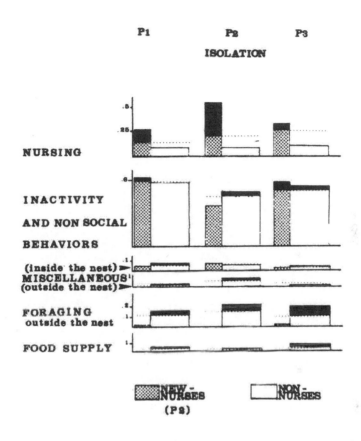

Figure 1.11.   Behavioral profiles of ants which
    took on the nursing role during isolation
    (new nurses) compared with those ants which
    did not take on this role (non-nurses).   The
    groups in period P1 (control) are based
    retrospectively on those established during
    P2. (Modified from Fresneau and Lachaud
    1985.)

1959; Hohorst 1972; Billen 1982).
    A study of age-related changes in behavior in
the genus Myrmica (Cammaerts-Tricot 1974) clearly
showed the existence of an ontogenesis of
physiological mechanisms leading the workers to
change their workplace from inside to outside the
nest.   Using progressive changes in cuticular
pigmentation, five age groups were identified (0-3

days, 3-16 days, 16-90 days, 90-180 days, and over 180 days). The author was able to show that the volume of secretions from the mandibular gland and from Dufour's gland increased with the age of the ant (these glands are implicated in alarm and recruitment behaviors). This change was paralleled by an increase in their responsiveness to these substances, as shown by defense reactions (Cammaerts-Tricot 1975). The youngest ants did not participate in nest defense; workers from the second age group were able to secrete trail signals; ants from the third and fourth age groups were heavily involved in recruitment behaviors. Workers from the fifth group--more than six months old--were involved in "the best form of defense": attack. However, this schema of age-related changes only has a statistical value, due to the global approach used in the study. If we examine the results closely, it appears that there is a strong inter-colony variation in this effect: overall, some individuals of all age groups responded in the different experimental procedures.

What then are the results when the analysis of behavior is based upon individual ants? Fresneau (1984) studied the social organization of Neoponera obscuricornis and then dissected the individuals in a given colony. He found that in this species social organization is characterized by the integration of the queens into the behavioral subcastes, with 5 behavioral subcastes in all. In the sacrificed colony, there were 7 wingless females and 57 workers. The structure was as follows:

1. Subcaste 1: Egg nurses
2. Subcaste 2: Nurses of larvae and pupae
3. Subcaste 3: Socially inactive ants
4. Subcaste 4: Intermediates (Domestics and Preforagers)
5. Subcaste 5: Foragers

A summary of the results of the analysis are given in the correlation diagram in Figure 1.12.
The link between the level of ovarian development and social behavior in females and in workers seems clear. The great majority of nurses

have developed ovaries. If we examine the behavior of individuals in relation to the total number of behavioral acts oriented towards the brood (on the right of the diagram), it is apparent that this "third dimension" of the figure shows a substantial inter-individual variability, especially in subcastes 2 and 3. This variation in the level of nursing is probably related to the variation in the functional maturation of the ovaries, for both workers and females. The existence of "atypical" individuals amongst both the egg nurses and the intermediate-preforagers suggests that there is a complex relationship between physiological maturation and the performance of behaviors which are socially integrated into the collective activity of the nest. Difficulties in controlling the age of members of the society make the researcher's job much more difficult and limit the conclusions we can draw. However, this kind of study will undoubtedly represent an extremely rich and important direction for future investigations.

ETHOGENESIS OF INDIVIDUAL BEHAVIOR

It is only recently that attention has been focused on the mechanisms involved in the development of behavior in the ant. One area of investigation has been the role of early processing of environmental stimuli as given behavioral patterns develop. Such studies help us to begin to understand how the behavioral orientation of an individual changes, based both on innate tendencies or predisposition and on individual history.

This type of experiment involves either isolating the ant, studying it in its social context, or combining these two methods.

Study of the Behavioral Potential of the Isolated Individual

Using this method we isolated 50 Formica polyctena Foerst workers two hours after the imaginal moult. We used individuals of the same

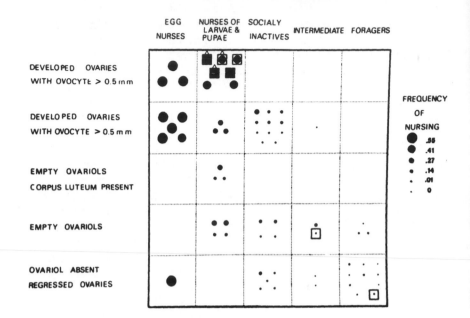

Figure 1.12. Graphic representation of the
relationship between social status
(behavioral subcaste), grouped in columns,
and five ovarian conditions, shown in rows.
The size of the symbol for each ant is
proportional to the frequency of nursing
behavior observed directed towards the
brood. (From Fresneau 1984.)

size (7 mm length) that had eclosed on the same
day from the same nest. Each was reared in the
same condition: the ant was placed in a glass
tube together with two depigmented pupae from the
same colony, and with abundant food. A behavioral
test was carried out on these individuals at five
times: at 1 hour, and at 1, 3, 6 and 11 days after
isolation. The behavioral test consisted of
tapping the tube for 20 seconds with a small pair
of tweezers, which induced the worker to transport
the pupae (the pupae were replaced after each
test). This age range was chosen because in F.
polyctena pupal tending behavior is at a maximum
during this period (Otto 1958). The first
transporting responses were found at the second

test period (1 day; 4 percent of individuals), and showed strong increases at 3 days (12 percent) and 6 days (18 percent). At the fifth test period, 22 percent of the individuals showed the characteristic carrying behavior (Figure 1.13). The most interesting result is the reliability of the test. Ten of the eleven individuals showing the carrying behavior at 11 days had already responded in one to three previous tests.

This highly significant result ($P < .0001$) shows that essentially identical ants--reared in the <u>same</u> manner, of the <u>same</u> size, born on the <u>same</u> day, coming from the <u>same</u> colony-- nevertheless show different individual potentials for a behavior of substantial social importance. Moreover, these different potentials were shown outside of the regulatory social pressure of the colony. This result suggests an interindividual variability in predisposition to certain behaviors.

The Role of Individual Experience During the Early Imaginal Stage

What happens when a previously isolated ant is reintroduced into its colony? This problem has recently been studied in the Neotropical ponerine ant <u>Ectatomma tuberculatum</u> (Champalbert 1985, 1986). The nursing behavior of such individuals reintroduced after 10 days of isolation at age 0, 2, 4 or 8 days, shows interesting results when compared with the behavior of control, non-isolated workers.

Non-isolated workers were studied individually in the society, using the systematic scanning method, as shown in Figure 1.14. They generally perform the first expressions of nursing behavior at 4 days old. The levels of this behavior increase progressively, with a maximum at about 20 days old. After this age there is a progressive decrease until about 40 days old. At this age, control ants show only a minimal level of nursing behavior. Ants isolated for 10 days following emergence show a similar profile after they have been reintroduced. However, the developmental change is shifted in time, as a

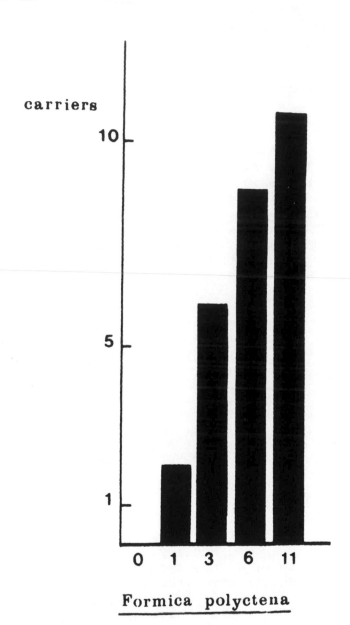

Figure 1.13. Cumulative responses of callow *Formica* *polyctena* workers submitted to five pupae-carrying tests during the first eleven days of life. Ordinate: the number of workers responding to the test. Abscissa: age of the ant at the test.

function of the period of isolation, as though the ant had "regressed" to one day old when replaced in a social context. When ants of 4 or 8 days old are isolated, this social behavior reappears soon after reintroduction. However, there are severe behavioral consequences in ants isolated at 2 days old. The development of nursing behaviors is substantially and definitively upset, although there are substantial interindividual

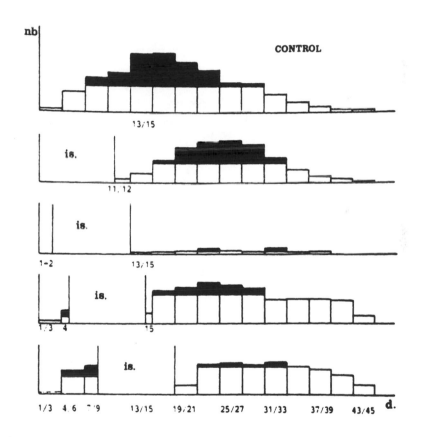

Figure 1.14. Consequence of a 10-day isolation period (at day 0, 2, 4 and 8 postemergence) on the total number of nursing behavioral acts (on ordinate). The black bars represent the nursing activity superior to the mean level for the group for the total duration of the experiment. The white bars represent the nursing activity inferior to this mean. (Modified from Champalbert 1985.)

variations. The reintroduced ant shows high levels of inactivity and a low level of nursing behavior towards the brood. The fact that isolation has such a severe effect only at age 2 days suggests that there is a complex sensitive period involved in the ethogenesis of the individual.

During this period social stimuli obviously play a vital part in the development of the individual's social role.

The Importance of Individual Experience During the Preimaginal Stage

Isingrini et al. (1985) have recently shown that the spontaneous preference for nestmate larvae observed soon after imaginal eclosion in the Mediterranean ant Cataglyphis cursor Fonscolombe is due to conditioning during the larval stage. Eggs of C. cursor were transferred from a parent colony to a recipient colony, producing larvae which had entirely developed in an alien, adoptive colony.

Immediately after pupation the cocoons were transferred back to the parent colony until eclosion (which occurred approximately 15 days afterwards). The newly-eclosed ants spent five days in the colony before being given a choice between nestmate larvae and larvae from the "alien" (adoptive and familiar) colony, where they lived their larval stage. Figure 1.15 shows that these ants preferred to tend larvae from the alien colony, whereas control ants always preferentially tended nestmate larvae. Furthermore, workers resulting from larvae adopted only at the third instar failed to prefer alien larvae. This means that the integration of the colony stimuli by an individual depends either on the duration of exposure or the age of the larvae. This first demonstration of preimaginal conditioning in social behavior shows that in C. cursor early spatial proximity is more important than genetic similarity for nestmate recognition. The history of the individual larva can thus influence the subsequent nursing behavior of the resultant worker. This ethogenetic process may be

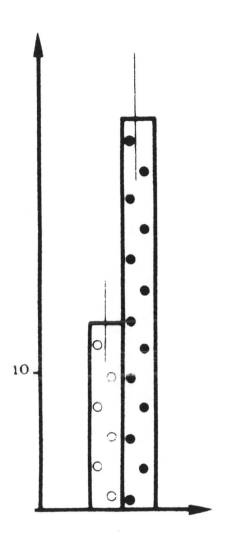

Figure 1.15. Mean number (and confidence counts)
    per adult ants of brood-nursing acts (licking
    and carrying) directed towards nestmate
    larvae (open circles) and alien larvae from
    the adoptive colony (filled circles).
    (Modified from Isingrini et al. 1985.)

important in other ant species, at least in the
Formicinae.

CONCLUSION

The study of the individual basis of social behavior in ants has only recently developed. Different methods have been used, which have various advantages and disadvantages. In general, they enable us to develop a more complete view of the factors affecting the behavioral development of the individual in its society. The main factors affecting the social behavior of individual ants are:

1. membership of a physical subcaste
2. individual sensitivity and responsivenss (idiosyncrasy)
3. personal history during ethogenesis (principally at larval and callow stages)
4. physiological modifications with age
5. state of the society

This last point--the state of the society--is important in evolutionary and developmental terms. The society has to be close to the equilibrium where individuals are distributed among all behavioral subcastes. The equilibrium is maintained by "régulation sociale" (Grassé 1946) or "social homeostasis" (Emerson 1956). This idea suggests a comparison between the cell and the individual ant, and leads us to consider that ants are members of a superorganism, as pointed out by Wheeler (1926). This theory is currently experiencing a revival (Lumsden 1982; Wilson 1985b; Jaisson 1985), and is increasingly supported by the study of individual social behavior.

As in other areas of biology, the study of social insects has been biased for a long time towards a "biology of the mean." Recent studies have clearly shown that a "socially average" individual does not exist, and that insect societies are plainly far more than the simple average (or sum) of their members.

Although it is difficult, the only way of escaping from this "fascination with the mean" is to study individual traits of social behavior and to try to conceptually integrate them into a global vision ot the society. That is, after all, the phylogenetic and ontogenetic context in which these traits have developed.

SUMMARY

Numerous studies have investigated the individual components of social behavior in ants. A critical review of the different approaches is made with special reference to the two main methods: sporadic sampling and systematic scanning. Many of these studies result in the description of species specific behavioral repertoires. Only some of them are able to identify the actual distribution of behavioral acts between the members of the society. Relationships between individual behavior and polymorphism, age, experience during ethogenesis, idiosyncrasy or physiological correlates are emphasized. The Discussion underlines the multifactorial origin of individual behavior and its integration to the social entity through a homeostatic, regulated, way.

ACKNOWLEDGMENTS

We are very grateful to Dr. M. Cobb for the English translation, to B. Corbara for his useful help in the preparation of figures and photographs, and to L. Simon and M. Giraud for the typing.

REFERENCES

Benzecri, J.P. 1973. L'analyse des données. 2 vols. Paris: Dunod.
Bertin, J. 1977. La graphique et le traitement graphique de l'information. Paris: Flammarion.
Billen, J. 1982. Ovariole development in workers of Formica sanguinea Latr. (Hymenoptera: Formicidae). Insectes Sociaux 29:86-94.

Brown, W.L., Jr. 1973. A comparison of the Hylean and Congo-West-African rain forest ant faunas. In Tropical forest ecosystems in Africa and South America: A comparative review, ed. B.J. Meggers, E.S. Ayensu and W.D. Duckworth, Washington: Smithsonian Institution Press. 161-185.

Calabi, P., Traniello, J.F.A., and Werner, M.H. 1983. Age polyethism: its occurrence in the ant Pheidole hortensis and some general considerations. Psyche 85:395-412.

Cammaerts-Tricot, M.C. 1974. Production and perception of attractive pheromones by differently aged workers of Myrmica rubra. Insectes Sociaux 21:235-248.

Cammaerts-Tricot, M.C. 1975. Ontogenesis of the defense reactions in the workers of Myrmica rubra. Animal Behaviour 23:124-130.

Champalbert, A. 1985. Ethogénese du comportement social et variabilité chez la jeune fourmi primitive Ectatomma tuberculatum. (Hymenoptera, Ponerinae). Thèse Doct. 3ème cycle, Université Paris XIII.

Champalbert, A. 1986. Individual ontogenesis of social behavior in Ectatomma tuberculatum (Ponerinae) ants. In The individual and the society, ed. L. Passera and J.P. Lachaud, 127-137. Toulouse: Privat.

Cole, B.J. 1980. Repertoire convergence in two mangrove ants, Zacryptocerus varians and Camponotus (Colobopsis) sp. Insectes Sociaux 27:265-275.

Corbara, B., Fresneau, D., Lachaud, J.P., Leclerc, Y., and Goodall, G. 1986. An automated photographic technique for behavioural investigations of social insects. Behavioural Processes 13:237-249.

Cordero, A.D. 1963. An unusual behavior of the leafcutting ant queen Acromyrmex octospinosa (Reich). Revista de Biologia Tropical 11:221-222.

Dejean, A. 1982. Quelques aspects de la prédation chez des fourmis de la tribu des Dacetini (Formicidae, Myrmicinae). Thèse Doct. d'Etat, Université de Toulouse.

Dobrzanska, J. 1959. Studies on the division of labour in ants, genus Formica. Acta Biologiae Experimentalis 19:57-81.

Emerson, A.E. 1956. Regenerative behavior and social homeostasis in termites. Ecology 37:248-258.

Fagen, R.M., and Goldman, R.N. 1977. Behavioral catalog analysis method. Animal Behaviour 25:261-274.

Fresneau, D. 1984. Développement ovarien et statut social chez une fourmi primitive Neoponera obscuricornis Emery (Hym., Formicidae, Ponerinae). Insectes Sociaux 31:387-402.

Fresneau, D., and Charpin, A. 1977. Une solution photographique au problème du marquage individuel des petits insectes. Annales de la Société Entomologique de France (N.S.) 13:1-5.

Fresneau, D., Garcia Perez, J., and Jaisson, P. 1982. Evolution of polyethism in ants: observational results and theories. In Social insects in the tropics, ed. P. Jaisson, vol. 1, 129-155. Paris: Presses de l'Université Paris XIII.

Fresneau, D., and Lachaud, J.P. 1985. La régulation sociale: données préliminaires sur les facteurs individuels contrôlant l'organisation des tâches chez Neoponera apicalis (Hym., Form., Ponerinae). Actes des Colloques Insectes Sociaux 2:185-193.

Fridman, S., and Avital, E. 1983. Foraging by queens of Cataglyphis bicolor nigra (Hymenoptera: Formicidae): an unusual phenomenon among the Formicinae. Israel Journal of Zoology 32:229-230.

Grassé, P.P. 1946. Sociétés animales et effet de groupe. Experientia 2:77-82.

Greenberg, L., Fletcher, D.J.C., and Vinson, S.B. 1985. Differences in worker size and mound distribution in monogynous and polygynous colonies of the fire ant Solenopsis invicta Buren. Journal of the Kansas Entomological Society 58:9-18.

48

Herbers, J.M. 1982. Queen number and colony
ergonomics in Leptothorax longispinosus. In
The biology of social insects, ed. M.D.
Breed, C.D. Michener, and H.E. Evans, 238–
242. Boulder: Westview Press.
Herbers, J.M. 1983. Social organization in
Leptothorax ants: within and between-species
patterns. Psyche 90:361–386.
Hohorst, B. 1972. Entwicklung und Ausbildung der
Ovarien bei Arbeiterinnen von Formica
(Serviformica) rufibarbis Fabricius
(Hymenoptera: Formicidae). Insectes Sociaux
19:389–402.
Isingrini, M., Lenoir, A., and Jaisson, P.
1985. Preimaginal learning as a basis of
colony-brood recognition in the ant
Cataglyphis cursor. Proceedings of the
National Academy of Sciences USA 82:8545–
8547.
Jaisson, P. 1985. Social behaviour. In
Comprehensive Insect Physiology, Biochemistry
and Pharmacology, ed. G.A. Kerkut and L.I.
Gilbert, vol. 9, 673–694. Oxford: Pergamon
Press.
Lachaud, J.P., and Fresneau, D. 1985. Les
premières étapes de l'ontogenèse de la
société chez Ectatomma tuberculatum et
Neoponera villosa (Hym., Form., Ponerinae).
Actes des Colloques Insectes Sociaux 2:195–
202.
Le Masne, G., and Bonavita, A. 1969. La
fondation des sociétés selon un type
archaique par une fourmi appartenant a une
sous-famille évoluée. Comptes Rendus de
l'Académie des Sciences, Paris 269:2373–2376.
Lenoir, A. 1979a. Le comportement alimentaire et
la division du travail chez la fourmi Lasius
niger. Bulletin Biologique de la France et
de la Belgique 113:79–314.
Lenoir, A. 1979b. Feeding behaviour in young
societies of the ant Tapinoma erraticum L.:
trophallaxis and polyethism. Insectes
Sociaux 26:19–37.
Lumsden, C.J. 1982. The social regulation of
physical caste: the superorganism revived.
Journal of Theoretical Biology 95:749–781.

McDonald, P., and Topoff, H. 1985. Social
    regulation of behavioral development in the
    ant, Novomessor albisetosus (Mayr). Journal
    of Comparative Psychology 99:3-14.
Morel, L. 1983. Contribution à l'étude des
    interactions sociales chez les jeunes
    ouvrières de la fourmi Camponotus vagus.
    Scop.: developpement du comportement
    trophallactique et régulation de
    l'agressivité. Thèse Doct. 3ème cycle,
    Université Aix-Marseille II.
Oster, G.F., and Wilson, E.O. 1978. Caste and
    ecology in the social insects. Princeton,
    New Jersey: Princeton University Press.
Otto, D. 1958. Über die Arbeitsteilung im Staate
    von Formica rufa rufo-pratensis minor Gössw.
    und ihre verhaltensphysiologischen
    Grundlagen: Ein Beitrag zur Biologie der
    roten Waldameise. Wissenschaftliche
    Abhandlungen der Deutsche Akademie der
    Landwirtschaftswissenschaften, Berlin 30:1-
    169.
Peeters, C.P. 1982. The reproductive strategy of
    the ponerine Ophthalmopone berthoudi: an
    insight into the evolution of ant
    eusociality. In The biology of the social
    insects, ed. M.D. Breed, C.D. Michener, and
    H.E. Evans, 220-221. Boulder, Colorado:
    Westview Press.
Perez-Bautista, M., Lachaud, J.P., and Fresneau,
    D. 1986. La division del trabajo en la
    hormiga primitiva Neoponera villosa
    (Hymenoptera: Formicidae). Folia
    Entomologica Mexicana 65:119-130.
Provost, E. 1983. Une nouvelle méthode de
    marquage permettant l'identification des
    membres d'une société de fourmis. Insectes
    Sociaux 30:255-258.
Ross, K.G., and Fletcher, D.J.C. 1985.
    Comparative study of genetic and social
    structure in two forms of the fire ant,
    Solenopsis invicta (Hymenoptera:
    Formicidae). Behavioral Ecology and
    Sociobiology 17:349-356.

Schmid-Hempel, P. 1983. Foraging ecology and colony structure in two sympatric species of desert ants, Cataglyphis bicolor and Cataglyphis albicans. Ph.D. diss., University of Zurich.

Schmid-Hempel, P., and Schmid-Hempel, R. 1984. Life duration and turnover of foragers in the ant Cataglyphis bicolor (Hymenoptera, Formicidae). Insectes Sociaux 31:345-360.

Smeeton, L. 1982. The effect of age on the production of reproductive eggs by workers of Myrmica rubra L. (Hym., Formicidae). Insectes Sociaux 29:465-476.

Verron, H., and Barreau, S. 1974. Une technique de marquage des insectes de petite taille. Bulletin Biologique de la France et de la Belgique 108:259-262.

Wehner, R., Harkness, R.D., and Schmid-Hempel, P. 1983. Foraging strategies in individually searching ants, Cataglyphis bicolor (Hymenoptera, Formicidae). In Information processing in animals, ed. M. Lindauer, vol. 1, 11-53. Akademie der Wissenschaften und der Literatur, Mainz. Stuttgart: Gustav Fischer Verlag.

Weir, J.S. 1958. Polyethism in workers of the ant Myrmica (Part I). Insectes Sociaux 5:97-128.

Wheeler, W.M. 1926. Les sociétés d'insectes. Leur origine. Leur evolution. Paris: Doin.

Wilson, E.O. 1968. The ergonomics of caste in the social insects. American Naturalist 102:41-66.

Wilson, E.O. 1971. The insect societies. Cambridge, Massachusetts: Harvard University Press.

Wilson, E.O. 1976a. A social ethogram of the neotropical arboreal ant Zacryptocerus varians (F. Smith). Animal Behaviour 24:354-363.

Wilson, E.O. 1976b. Behavioral discretization and the number of castes in an ant species. Behavioral Ecology and Sociobiology 1:141-154.

Wilson, E.O. 1983. Caste and division of labor in leaf-cutter ants (Hymenoptera: Formicidae: Atta). IV. Colony ontogeny of A. cephalotes. Behavioral Ecology and Sociobiology 14:55-60.

Wilson, E.O. 1984. The relation between caste ratios and division of labor in the ant genus Pheidole (Hymenoptera: Formicidae). Behavioral Ecology and Sociobiology 16:89-98.

Wilson, E.O. 1985a. The principles of caste evolution. Fortschritte der Zoologie 31:307-324.

Wilson, E.O. 1985b. The sociogenesis of insect colonies. Science 228:1489-1495.

Wilson, E.O. 1986. Caste and division of labor in Erebomyrma, a genus of dimorphic ants (Hymenoptera: Formicidae: Myrmicinae). Insectes Sociaux 33:59-69.

Wilson, E.O., and Hölldobler, B. 1985. Caste-specific techniques of defense in the polymorphic ant Pheidole embolopyx (Hymenoptera: Formicidae). Insectes Sociaux 32:3-22.

Wilson, E.O., and Hölldobler, B. 1986. Ecology and behavior of the Neotropical cryptobiotic ant Basiceros manni (Hymenoptera: Formicidae: Basicerotini). Insectes Sociaux 33:70-84.

2

The Group Context in
Role Switching in Harvester Ants

Deborah Gordon

Social insects live in groups. Every aspect of their biology--reproduction, obtaining resources, modifying the environment--takes place at the group level as well as at the individual level. The behavior of individuals is embedded in the behavior of groups. To explain individual behavior, it is essential to take into account the ways in which every individual's behavior is determined by the group context. Current research on the behavior of social insects is producing important new insights about individual behavior. To interpret these results, we will need to know more about colony organization at the group level. How does individual behavior depend on colony needs? Or, to turn the question around, how do changes in colony needs affect what individuals do?

This question has not yet received much attention. Instead, most of the questions posed by current research concern the behavior of individuals. We ask, what task does an individual do? What physiological events determine this task? Does it depend on age, size, hormonal processes, food intake? How likely is this individual to switch tasks?

The answers to these questions depend to some extent on factors at the individual level. Not all individuals are alike; some are more likely to behave in a particular way than are others. Individuals change their behavior in response to changed conditions. This response depends on the individual: some ants might respond to

disturbance in the nest by carrying out brood;
others respond by circling around and alerting
their nestmates.

New questions arise when we consider how an
individual's behavior is also affected by factors
at the group level. How does an individual's
behavior depend on colony-level allocation of
workers into different tasks? Colonies respond to
changes in their environments in species-typical
ways. For example, some species will respond to
disturbance by sending large numbers of aggressive
individuals out towards it; others will retreat
further into their nests. Colonies modulate
foraging and brood production in response to
changes in food supply, nest construction in
response to changes in weather, territorial
behavior in response to changes in the pattern of
encounters with neighboring species. It is clear
that this modulation is predictable and rule-
governed, though we don't yet know the rules.

These rules for how the colony modulates its
behavior in response to environmental change are
the dynamics of colony behavior. One way the
colony responds to environmental change is to
modify the allocation of workers into different
tasks. The dynamics of this allocation are the
rules for how the numbers of workers in different
tasks change as the environment changes. The
probability that a worker will switch from task A
to task B differs across individuals. But whether
a worker actually switches tasks or not also
depends on how the colony is channeling workers
into different tasks in response to changed
conditions. Individual behavior is constrained by
the dynamics of colony behavior. We cannot
interpret data on individual differences except in
the context of these dynamics.

Investigation of the dynamics of colony
behavior is just beginning. Seeley (1986) found
that the number of workers a honeybee colony
allocates to foraging outside the nest depends on
fluctuations in the amount and quality of incoming
nectar. Whether an individual honeybee will go
out to forage depends on attributes of that
individual, such as its age and responsiveness to
the stimuli signaling a need for food. Those
stimuli, however, are regulated by the dynamics of

group behavior. Predicting what a particular forager will do entails predicting how changes in the colony's need for food affect the numbers of individuals recruited to forage.

Seeley's work shows that the number of workers doing one task (foraging) depends on environmental change (in food supply). In my own work on harvester ants, I have examined a further question about the dynamics of group behavior. Do environmental changes that affect the numbers engaged in one task also affect the numbers engaged in other tasks?

I have begun a series of perturbation experiments, done in the field in southeastern Arizona, using colonies of the red harvester ant, Pogonomyrmex barbatus (F. Smith). I caused changes in the numbers foraging, doing nest maintenance work, or patrolling (Gordon 1986, 1987). I interfered with foraging by putting out small barriers on the foraging trails. Foragers eventually went around or over them, but the overall effect was a decrease in numbers foraging. I caused an increase in nest maintenance by putting out piles of toothpicks early in the morning when nest maintenance workers are beginning their activity. They moved all the toothpicks to the edge of the nest mound, where they were then ignored, before these workers resumed their usual tasks.

The experiments directly change the numbers of ants engaged in a particular task. Putting out barriers on the foraging trails causes a decrease in numbers foraging. A pile of toothpicks near the nest entrance causes the number of ants doing nest maintenance work to increase. The experiments also have a striking indirect effect. When the number performing one task changed, the numbers in other tasks changed as well. When foraging was decreased, nest maintenance increased. When nest maintenance was increased, foraging decreased. Since the exterior work force comprises only 10-20 percent of the colony (MacKay 1981), there is a large pool of reserve workers inside the nest. Indirect effects on the numbers engaged in tasks that were not perturbed should not be assumed to be the result of the allocation of a fixed total number of workers.

These results show that the number doing one task depends on the number doing another. For example, the colony's allocation into foragers does not depend merely on events affecting foragers. It also depends on the number doing nest maintenance work. The numbers doing different tasks are interdependent. To predict the behavior of an individual forager, we need to know not only how the number of foragers depends on food supply, but how it depends on the numbers doing other tasks as well.

This result implies, then, a set of numerical relationships between different worker groups. Further results demonstrate that these numerical relationships are themselves modulated by environmental change. When I interfered with foraging in one group of colonies and nest maintenance in a second group, I obtained the results described above. Events affecting one group of workers cause changes in the numbers doing other tasks. In a third group of colonies I interfered with both foraging and nest maintenance at once. Results were very different from the additive effect of summing the two single perturbations.

The relative priorities of different activities depend on the level of stress the colony is subjected to. Combined perturbations show synergistic effects. Perturbations that decreased foraging led to relative increases in nest maintenance; those that caused an increase in nest maintenance led to relatively decreased foraging. Thus both single perturbations had the same result: increased nest maintenance and decreased foraging. Combining both perturbations --subjecting a colony to both barriers and toothpicks at once--had the opposite result. Relative to the summed effects of the single perturbations, foraging increased and nest maintenance decreased. This shows that the rules for how number in one activity depends on number in another, themselves change with conditions. With single perturbations, nest maintenance and foraging are of reciprocal priority, and it can go either way. When more than one activity is perturbed, foraging is of higher priority than nest maintenance.

Not only do numbers doing one task change in response to changed numbers in other tasks, but this response is itself modulated by the amount of perturbation. Coming back to the individual level, it is clear the group factors affecting an individual are complex. An individual's behavior depends on the kind of individual it is, on environmental events affecting the group it is participating in, on events that affect groups engaged in other tasks, and on which combination of other groups is affected.

These dynamics change as a colony matures. The way younger colonies modify the allocation of workers in response to environmental change is different from older colonies. Responses to the increased stress of combined perturbations depend on colony age. For example, the older colonies' reaction to combined perturbations shows a much stronger emphasis on foraging than does the reaction of younger colonies, which tend to increase patrolling more under increased stress.

These results raise many questions about the dynamics of group behavior at the colony level. How do the rules for the modulated allocation into different tasks, depending on environmental change, develop? How do these rules affect colony fitness? How does interspecific variation in colony flexibility affect the ecological structure of communities? At the individual level there are many other interesting questions. How do workers doing one task assess the number doing another? Do the signals depend merely on the number of workers doing a task or, as in the case of honeybee foragers described above, on the amount of each task that gets done?

Another interesting question at the individual level concerns the mechanism for changes in number doing a particular task. Do workers switch tasks or are they recruited from the reserves inside the nest? Experiments in which workers were removed (Gordon 1986) suggest that in P. barbatus new nest maintenance workers and foragers are recruited from a common pool. Further experiments with marked individuals are in progress to explore this question.

The question of how individuals respond to environmentally-induced changes in the number of workers doing other tasks links the investigation of individual differences with the dynamics of group behavior. To predict what an individual will do, we need to keep three questions in mind. First, how do that individual's attributes (such as age or size) determine its response to environmental change? For example, Meudec and Lenoir (1982) found that certain active individuals will respond to disturbances by moving brood, while in the same situation other inactive ones will not. Second, how do all the individuals doing some task respond to environmental changes related to their task? For example, harvester ants will decrease the number foraging in response to barriers on the foraging trail; honeybee colonies will respond to changes in food supply by modulating the number of foragers. Third, how do the individuals doing some task respond to events causing changes in the numbers doing other tasks? It is not possible to understand answers to the first of these questions without also considering the next two. To interpret individual differences we must also investigate the dynamics of group behavior.

SUMMARY

A social insect colony adjusts its behavior in response to changes in its environment. These adjustments take place at both the individual and group levels. At the individual level, workers change tasks or modulate the intensity of activity to meet colony needs. At the group level, the size of a group engaged in some task, and the time and duration of that worker group's activity, are adjusted in response to colony needs. In harvester ant colonies, these adjustments at the group level arise out of complex interactions among different groups engaged in various tasks. Such interactions between worker groups change as the colony matures. To understand how a colony's social organization responds to the colony's environment, we must begin to consider the

dynamics of group, as well as individual,
behavior.

REFERENCES

Gordon, D.M. 1986. The dynamics of the daily
round of the harvester ant colony
(Pogonomyrmex barbatus). Animal Behaviour
34:1402-1419.
Gordon, D.M. 1987. Group-level dynamics in
harvester ants: young colonies and the role of
patrolling. Animal Behaviour 35:833-843.
MacKay, W.P. 1981. A comparison of the nest
phenologies of three species of harvester ants
(Hymenoptera: Formicidae). Psyche 88:25-74.
Meudec, M., and Lenoir, A. 1982. Social
responses to variation in food supply and nest
suitability in ants (Tapinoma erraticum).
Animal Behaviour 30:284-301.
Seeley, T.D. 1986. Social foraging by honeybees:
how colonies allocate foragers among patches
of flowers. Behavioral Ecology and
Sociobiology 19:343-354.

3

Interindividual Differences Based on Behavior Transition Probabilities in the Ant *Camponotus sericeiventris*

Prassede Calabi and Rebeca Rosengaus

INTRODUCTION

Division of labor among morphological and/or age classes is ubiquitous in social insects (Michener 1974; Wilson 1971). As a result, earlier studies of behavior and division of labor only considered each age and/or physical caste as a whole. That consideration resulted in caste-typical averages which smoothed over individual variability and gave the impression that caste or colony responses were deterministic. However, recent studies of individually marked workers and experimental manipulations of caste ratios have shown that there is considerable individual variability and that caste responses are not deterministic (for example, this volume, Calabi [1986 and references; in press], Wilson [1985 and references], Mirenda and Vinson [1981]). We here present data from 12 known-age and 68 other individually marked workers of the ant Camponotus sericeiventris Guerin that show differences among individuals and between cohorts with respect to the relative frequencies of both behaviors and behavior pairs. In addition, we address the question of roles, or sets of behaviors "linked together by relatively high transition probabilities" (Oster and Wilson 1978:122). Specifically, we ask which role(s) can be defined for the minor worker physical caste of C. sericeiventris, and whether a role is the property of an individual or of a whole caste. We also

consider applications of behavior pair transition
analysis.

MATERIALS AND METHODS

The Colony

The study was conducted on a single colony of
the ant Camponotus sericeiventris started from a
founding queen collected on Barro Colorado Island,
Panama, in 1979. At the time of this study the
colony numbered approximately 2500 adult
workers. Observations were carried out on four
subcolonies, each with a fixed physical caste
ratio of both adult and brood classes, and
approximately constant caste ratios between inside
and outside workers of each of the three adult
physical castes. (It is believed that an
association exists between age and the physical
location of adult workers: workers active inside
the nest tubes are likely to be young, while those
outside are likely to be old [Otto 1958].
However, since our data from known-age individuals
show that this association is indeed not fixed, we
do not consider locale as an index of age, but use
it simply as another aspect of colony
composition.) Each subcolony contained 100 adult
workers. The physical caste ratios used were
those observed in the whole colony during a
previous study (Busher et al. 1985): majors 10
percent, medias 4 percent, and minors 86
percent. The subcolonies were also provided with
20 pupae and 50 larvae of mixed sizes, and
approximately 4 cc of eggs and microlarvae.
Subcolonies were censused weekly and restocked
from the main colony as necessary to maintain
those proportions.

Except during the actual hours of
observations, the subcolonies were kept in plastic
boxes (23 x 32 x 10 cm) within the round foraging
arena of the main colony (diameter 120 cm). The
plastic boxes housing subcolonies had 8 openings
of 4 cm covered with a wire mesh sufficiently
large to allow antennal contact and trophallaxis
between workers inside the box (that is, from the

subcolony) and those outside it (from the main colony). Such exchanges, combined with the physical proximity of the subcolony to the main nest chamber and the queen, apparently permitted normal circulation of colony pheromones: ovarian inhibition of majors in the subcolonies was not interrupted, and behavior patterns by subcolony workers were like those in the main colony during the previous study (Busher et al. 1985). The main colony and subcolonies were provided with test tubes (15 x 2.3 cm) filled with water behind a cotton plug. All were fed freshly killed Tenebrio larvae and synthetic diet (Bhatkar and Whitcomb 1970). Subcolonies were fed during observation periods, when their nest box was removed from the main nest area and connected by a bridge to a special foraging arena in which the food was placed. This ensured also that leaving the nest box and foraging were unmistakable behaviors which could be unambiguously identified. Colonies were kept on a light cycle of 14:10 hours light:dark; most observations took place between 0700 and 2000 h.

Data Collection

Minor workers perform most of the quotidian tasks of the colony (Busher et al. 1985), and were thus the focus of this study. Individuals were uniquely marked on the thorax and gaster with Testor's® paints and were observed continuously for one or more half-hour periods, during which every behavior performed by the focal worker was entered directly into a computer via an ethogram program (Gladstein and Carlin, personal communication). A total of 80 ants from two groups were observed. The first group comprised ants of unknown age but characterized by their physical location in the subcolonies, either inside (Innendienst) or outside (Aussendienst [Otto 1958]) the nest tubes. We observed 35 inside and 33 outside workers for 25 and 20 hours respectively, during which over 4400 behaviors were recorded. Five additional workers were ultimately excluded from these analyses because they performed tasks in both locales. A second

set of 12 workers was individually marked upon
eclosion, all within a four-day period. Thus
these ants were a true age cohort, and were
subject to identical conditions at any age. Each
of these known-age workers was observed for an
average of two hours during the first month of
life; during their second, third, and (for the
five that survived) fifth months, each was
observed for an average of three hours per
month. A total of 120 hours of observations and
7996 behaviors were recorded for known-age,
individually marked workers.

Although a few additional behaviors were
recorded rarely, 18 behaviors were used in the
analyses. For some analyses those were collapsed
into 8 or 10 functional categories (Table 3.1).
Since ergonomic considerations are principal in
any study of caste and division of labor, the
behaviors were also divided into task and nontask
behaviors (sensu Oster and Wilson [1978]; see also
Calabi et al. [1983]. "Task" is used to denote a
set of acts which achieve some purpose of the
colony. Thus, although all tasks are behaviors,
not all behaviors are tasks.)

A behavior pair was defined as any consecutive
pair of different behaviors: thus by definition a
behavior could not "follow" itself. Although this
is not how others have defined transition (for
example, Herbers and Cunningham [1983]), it seemed
functionally more useful. This definition allowed
distinctions between frequency and duration of
behaviors, it measured actual frequencies with
which pairs of behaviors co-occurred, and it
allowed comparisons between such observed co-
occurrences and expected co-occurrences calculated
from the independent frequencies of the two
behaviors.

Analyses

Data were analyzed in several ways. The
twelve known-age ants were compared by relative
frequencies both of common behaviors (Table 3.3)
and of behavior pairs (Tables 3.2, 3.4, Appendix
3.1). (Use of relative frequencies or proportions
controls for differences among ants in number of

Table 3.1
Behaviors used in analyses

TASK BEHAVIORS

Brood Care

Groom egg - LE
Groom larva - LL
Groom pupa - LP
Handle egg - HNE
Handle larva - HNL
Handle pupa - HNP
Trophallaxis with larva - RGL

Nest Maintenance

Handle Nest Material - HNM
Lick Nest - LN

Trophallaxis - REC
Eat solid food - ESF
Eat liquid food - ELF

Groom adult - GRM

Forage - FOR

NONTASK BEHAVIORS

Locomote - LOC

Rest - RST

Selfgroom - SG

---

hours of observation, and allows comparisons
despite differences in activity rates [Calabi
1986].) Observed frequencies of behaviors were
compared with an expected frequency which was the
group mean. Behavior pair frequencies for each
individual were also compared to the group mean
using the normal approximation to a binomial test

Table 3.2
Data for known-age ants

	Ant Number				
	1	2	3	4	5
Age at[a]					
Death	174	>160	>100 <150	43	172
Foraging first seen	63	41	61	–	<30
Brood care first seen	<10	–	–	<10	<10
Behavior Pair[b]					
Brood-Brood	0.04	0	0	0.13	0
Brood-Task	0.04	0	0	0.01	0
Brood-Nontask	0.04	0	0	0.32	0.007
Groom-Task	0.01	0	0.005	0.005	0.03
Groom-Nontask	0.05	0.01	0.04	0.06	0.18
Trophallaxis-Task	0.01	0.03	0.02	0	0.03
Trophallaxis-Nontask	0.06	0.10	0.16	0.06	0.14
Forage-Task	0.01	0	0.01	0	0
Forage-Nontask	0.23	0.02	0.24	0	0.21
Nontask-Nontask	0.49	0.84	0.52	0.41	0.45
Task-Nontask	0.15	0 03	0.03	0.14	0.06
Sample Sizes					
Behavior Pairs	483	354	200	140	404
Behaviors	965	707	399	279	807
Tasks	299	64	100	98	234
Rates[c]					
Behaviors	32	44	36	46	38
Tasks	10	4	9	16	11
Percent Tasks	0.31	0.09	0.25	0.35	0.29

[a]Age in days. Dash (–) indicates that the behavior was never seen.

Table 3.2 (continued)

			Ant Number				
6	7	8	9	10	11	12	Group Mean
>100	>100	167	>100	>100	222	73	
<150	<150		<150	<150			
71	<30	61	<30	41	64	<30	
<10	–	–	–	–	–	<10	
0.06	0	0	0	0	0	0.02	0.02
0.01	0	0	0	0	0	0.01	0.002
0.15	0	0	0	0	0	0.03	0.04
0.01	0	0	0	0	0.01	0.02	0.01
0.05	0.03	0.10	0.03	0.04	0.08	0.05	0.07
0.01	0.01	0.01	0.01	0.02	0.02	0.13	0.02
0.09	0.09	0.18	0.17	0.19	0.14	0.05	0.15
0.01	0.01	0.004	0.003	0.01	0.01	0.26	0.007
0.24	0.51	0.19	0.38	0.30	0.17	0.02	0.21
0.49	0.36	0.50	0.40	0.44	0.57	0.41	0.47
0.15	0.02	0.01	0.02	0.03	0.04	0.14	0.05
483	225	263	282	404	425	377	
965	449	525	563	807	849	753	
299	144	135	171	239	169	196	
32	50	66	63	42	37	94	50
10	16	17	19	13	7	24	14
0.31	0.32	0.26	0.30	0.29	0.20	0.26	0.27

[b]Data given as relative frequencies.
[c]Rates given as number of behaviors per half hour.

Table 3.3
Differences per month in the relative frequencies with which known-aged ants performed each behavior

Behavior	Month 1	Month 2	Month 4	Month 5	All Months[b]
Brood Care	0.06 0.09 c	0.001 0.005 c	0.023 0.080 c	0.0 NS	0.03 0.07 c
Trophallaxis	0.11 0.16 c	0.10 0.19 NS	0.08 0.11 c	0.11 0.50 NS	0.01 0.15 c
Groom	0.06 0.10 NS	0.04 0.07 NS	0.02 0.03 NS	0.05 0.03 NS	0.04 0.08 NS
"Out"	0.07 0.12 NS	0.16 0.16 c	0.35 0.31 c	0.34 0.27 c	0.20 0.24 NS
Selfgroom	0.15 0.08 c	0.13 0.06 NS	0.18 0.04 NS	0.16 0.04 c	0.15 0.06 NS
Locomote	0.20 0.08 c	0.21 0.12 c	0.13 0.16 c	0.22 0.15 c	0.18 0.12 c
Rest	0.34 0.20 c	0.35 0.12 c	0.31 0.18 c	0.32 0.15 NS	0.34 0.16 NS

[a] At month 1 there were 12 ants; by month 5 only five ants still survived.
[b] Summary data for all ants across all months.
c = $P < .05$, NS = not significant.

(Zar 1974). The same test was used to compare behavior pair frequencies of inside and outside workers (Table 3.5). Finally, observed and expected frequencies of given behavior pairs, based on data from all workers combined, were compared via x^2 test (Tables 3.6 and 3.7), with the expected frequency of pair occurrence calculated via the following formula, based on the independent frequencies of the two behaviors:

$$Exp = \frac{A}{Tot} \times \frac{B}{Tot-A} + \frac{B}{Tot} \times \frac{A}{Tot-B},$$

A = observed frequency of behavior A

B = observed frequency of behavior B

Tot = sum of all behaviors.

RESULTS

Known-Age Individuals

The known-age ants differed with respect to whether a behavior was performed (and if so, at what age) and in the proportionate frequency of particular behavior pairs. Table 3.2 shows differences among the ants. They varied in the age at which they were first seen foraging (row 2): approximately 45 ± 20 days, excluding #4, which never left the nest box and died at day 43. These differences are not likely due to differences in longevity. Aside from #4, #11 died at 222 days, and the rest died between days 130 and 175 (row 1). The ants also varied with respect to whether and when they started brood care behaviors--either within 10 days of eclosion (5 ants), or never (7 ants) (row 3). Most compelling in terms of ergonomic considerations, one-third of the ants differed significantly from average in the number of tasks (row 19) or behaviors (row 18) performed per half hour. The mean plus or minus one standard deviation of different tasks per half hour was 14 ± 6 (or 27 ± 7 percent). By x^2 test, one worker (#12)

Table 3.4
Frequencies of performance of behavior pairs relative to the group average[a]

Behavior Pair	Ant Number												Totals	
	1	2	3	4	5	6	7	8	9	10	11	12	No. of ants $> \bar{x}$	No. of ants $< \bar{x}$
Brood–Brood	>	<	<	>	<	>	<	<	<	<	<	=	3	8
Brood–Task	>	<	<	>	<	>	<	<	<	<	<	=	3	8
Brood–Nontask	=	<	<	>	<	>	<	<	<	<	<	=	2	8
Groom–Task	=	<	<	<	>	<	<	<	<	<	=	>	2	8
Groom–Nontask	=	<	=	=	>	=	<	>	<	<	=	<	2	5
Troph.–Task	=	>	=	>	>	=	=	=	=	=	=	>	3	1
Troph.–Nontask	<	<	=	<	=	<	<	>	=	>	=	<	2	6

Table 3.4 (concluded)

Out-Task	=	=	=	=	=	=	=	=	=	=	=	=	>	1	—
Out-Nontask	=	<	>	<	=	>	>	<	>	=	>	=	<	3	3
Nontask-Nontask	=	>	<	<	=	<	<	=	>	>	<	>	<	3	5
Tasks per half hour	=	<	<	=	=	=	=	=	<	>	<	>	>	1	3

ᵃIndividual performed a given behavior pair more often than (>), less often than (<), or as often as (=) the group average.

Table 3.5
Differences in behavior pair transition frequencies between inside (Innendienst) and outside (Aussendienst) workers

Behavior Pair	Inside	Outside	Significance of Difference
Brood Care-Brood Care	0.05	0	b
Brood Care-Task	0.01	0	b
Brood Care-Nontask	0.22	0	b
Groom-Task	0.001	0.01	a
Groom-Nontask	0.13	0.02	a
Trophallaxis-Task	0.01	0.14	a
Trophallaxis-Nontask	0.14	0.08	a
Eat Liquid Food-Task	0	0.03	b
Eat Liquid Food-Nontask	0	0.01	b
Forage-Task	0.004	0.18	a
Forage-Nontask	0	0.56	b
Nontask-Nontask	0.38	0.13	a
Task-Nontask	0.54	0.66	a

No. of behavior pairs	1681	1749	NS
Hours of observation	25	20	NS
Number of ants	35	33	NS

[a]Statistically significant differences at $P < .05$; [b]Significance likely but not calculated because of zero values; NS = not significant.

Table 3.6
Expected and observed values for behavior pairs[a]

Behavior Pair	Observed	Expected	Significance of Difference
Brood-Task	91	101	NS
Brood-Nontask	370	772	c
Forage-Task	315	292	NS
Forage-Nontask	979	697	c
Groom-Task	34	28	NS
Groom-Nontask	253	360	c
Trophallaxis-Task	254	298	b
Trophallaxis-Nontask	375	301	c

[a]Based on independent occurrences among 68 ants.

[b] = $P < .05$; [c] = $P < .001$; NS = not significant (x^2 test).

performed significantly more than the average number of different tasks per half hour (24), and three ants performed significantly fewer (#2, #3, and #11, with 4, 7, and 9 tasks per half hour, respectively). Similarly, the mean ± one SD (standard deviation) of behaviors per half hour was 50 ± 17. Two ants (#8 and #12) changed behaviors significantly more often (66 and 94, respectively), and two (#1 and #3) changed less often than average (32 and 36). However, although a third of the ants differed from average behaviorally, an ant differing with respect to the number of tasks does not necessarily differ in number of behaviors, and vice-versa. Number 8 changed behaviors frequently (66 per half hour), but she did not perform tasks at above average rates (17/half hour). Number 1 changed behaviors

Table 3.7
Observed and expected behavior pair frequencies
among brood care behaviors, and between some
nontask and brood care behaviors

Behavior Pair[a]	Relative Frequency[b]	Significance (x^2)[c]	Observed/ Expected[d]
LE-GRM	=	–	2/2
LE-HE	>	104.0	34/7
LE-LL	<	9.9	4/17
LE-LOC	>	8.2	62/43
LE-LP	<	27.03	1/29
LE-RGL	<	–	0/3
LE-RST	>	17.5	20/8
LE-SG	>	8.0	25/14
HE-LE	>	104.1	34/7
HE-LOC	<	4.1	5/12
HE-LP	<	6.1	1/8
LL-GRM	>	–	5/1
LL-HL	>	–	9/2
LL-LE	<	9.9	4/17
LL-LP	<	4.5	9/18
LL-RGL	>	–	14/2
LL-RST	>	5.0	10/5
LL-SG	>	69.4	34/9
HL-LL	>	–	9/2
HL-LOC	>	14.4	22/10
HL-LP	<	–	1/4
LP-GRM	>	–	6/2
LP-HP	>	–	7/3
LP-LL	<	4.5	9/18
LP-LOC	>	151.7	89/26
LP-RST	>	7.2	17/9
LP-SG	>	9.2	28/16
LP-RGL	<	–	1/3
HP-LOC	>	9.8	12/5
HP-LP	>	–	7/3

(continued)

76

Table 3.7 (concluded)

Behavior Pair[a]	Relative Frequency[b]	Significance (χ^2)[c]	Observed/ Expected[d]
RGL-GRM	>	-	1/10
RGL-LOC	<	-	1/4
RGL-RST	=	-	1/1
RGL-SG	=	-	1/1

[a]HE=handle egg, HL=handle larva, HP=handle pupa, GRM=groom, LE=lick egg, LL=lick larva, LOC=locomote, LP=lick pupa, RGL=regurgitate to larva, RST=rest, SG=selfgroom.

[b]Relative frequency indicates whether a given behavior pair occurred more often than (>), as often as (=), or less often than (<) expected based on the independent occurrences of the two behaviors involved (see Methods for calculating the expected values).

[c]χ^2 values are given for pairs with adequate sample sizes; values >3.84 denote P <.05.

[d]Actual behavior pair frequencies.

less often than average (32/half hour), but she did not perform tasks at below average (10/half hour).

Table 3.3 shows differences per month and across all months in the relative frequencies with which individuals performed the behaviors named. The number of individuals differing from average changes by month (compare Figure 3.1), which is noteworthy, given that the ants are of the same age and are in a subcolony of constant physical caste ratios and under constant conditions. Frequency histograms for several common behaviors for five long-lived ants are presented in Figure 3.1.

Individuals differed also with respect to the proportionate frequencies of various behavior

pairs (Tables 3.2 and 3.4). The last column in
Table 3.2 is the average relative frequency for
all 12 ants combined; the last two columns in
Table 3.4 show how many ants performed particular
behavior pairs significantly more or less often
than expected, based on that average.
Unfortunately, more fine-grained analyses were not
possible for these data. When behavior pair
frequencies among all possible behavior
combinations were examined, sample sizes became
too small. ($\underline{N}$ = 300-600 behavior pairs per
individual ant, of which an average 78 percent
were non-task behaviors. Brood-brood transitions
are not in contradiction of our definition as
involving co-occurrence of different behaviors:
these are sums of transition frequencies among the
seven brood care behaviors.)

Inside and Outside Workers

Data from the 68 inside and outside workers
showed that, among workers from each locale, some
individuals performed no task behaviors. In
addition, there were significant differences by
group in behavior pair occurrences; task behaviors
tended to be clumped temporally; and brood care
behavior occurred in clusters by brood class.
Fourteen of 35 inside workers (40 percent)
performed no brood care during the average 43
minutes of observation per individual. Similarly,
18 percent of 33 outside workers were not seen to
forage during the average 36 minutes of
observation.
Since inside and outside workers perform
different behaviors, it is not suprising that the
two groups differ significantly in their
proportionate behavior pair frequencies (Table
3.5), based on the summed data of all 68.
However, even behavior pairs performed by both
groups can occur in different frequencies; for
example, insiders performed proportionately more
groom-nontask and trophallaxis-nontask pairs than
did outsiders, while the latter performed more
trophallaxis-task pairs.
Moreover, task behaviors tended to occur in
clusters. Although pairs of task behaviors co-

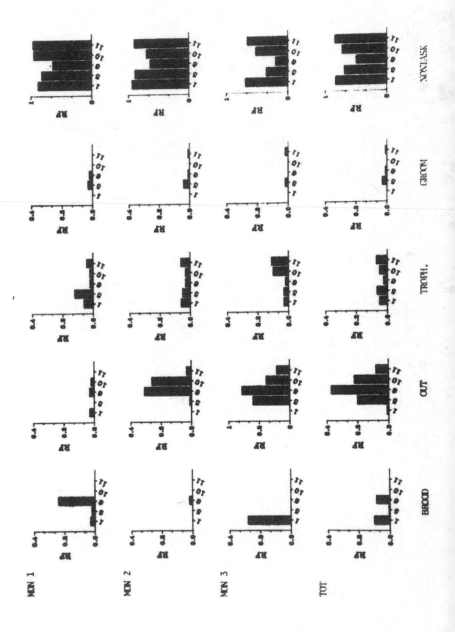

Figure 3.1. Relative frequencies of four task and all non-task behaviors for five ants. M1, M2, and M3 refer to months 1, 2, and 3, respectively; TOT is the sum across all months. Individual ants are identified along the x-axis. The y-axis is relative frequency. Note that the y-axis scale is constant except for OUT in M3, and for all months of nontask behaviors. Brood = all brood care, Troph. = trophallaxis, Out = forage plus leaving the nest box, Nontask = locomotion, nest, and selfgroom.

occurred as often as expected based on their
independent frequencies, transitions from those
groups to nontask behaviors occurred less often
than expected for three of four groups (Table
3.6).

Behavior pair frequencies among brood care
behaviors suggest that they tend to be temporally
grouped by brood class, although individual
workers do not specialize in caring for single
brood classes. Several behaviors directed at a
given brood class co-occur, but they are likely to
be followed by behaviors directed at a different
brood class. Thus, pairs of larva-care behaviors
co-occurred more often than expected, while larva-
care behaviors co-occurred with either egg- or
pupal-care behaviors less often than expected
(Table 3.7, Appendix 3.1). However, of 21 workers
performing brood care, 76 percent cared for more
than one brood class within a half hour
observation period.

Particular brood care behaviors also co-
occurred with non-task behaviors (selfgroom,
locomote, rest) more often than expected,
suggesting that although brood care behaviors as a
group occur in clusters, no particular behavior
ends or begins such a cluster. In other words,
there is no apparent sequence of brood care
behaviors within such a cluster.

DISCUSSION

Same-age individual ant workers differed
significantly with respect to proportionate
frequencies of both behaviors and behavior pairs,
as well as with respect to whether and at what age
they perform particular behaviors. We suggest
that these differences are due to different
sensitivities or thresholds among individuals to
work needs in the colony. It is particularly
noteworthy that these differences are qualitative
and not merely quantitative. Not only do
individuals differ at any age with respect to
proportionate performance of behaviors, they
differ absolutely with respect to whether those
behaviors are performed at all. In addition, the
data raise several interesting considerations

about ergonomics and division of labor,
particularly about age-based change in activity
rate, and factors affecting individual behavioral
ontogeny.

Failure by 58 percent of the 12 ants to show
brood care behaviors is striking, and makes more
credible, despite brief per-individual observation
periods, the failure by 40 percent of the unknown-
age inside workers to show brood care. Based on
flexibility experiments with other species (Calabi
1986; Wilson 1983), it is highly unlikely that
those workers were in any way incapable of brood
care. They simply failed to perform any. We
suggest that such individuals have higher
thresholds (lower sensitivities) to brood needs
than do active nurses. Thus, any colony would
have a mix of workers of varying sensitivities.
More sensitive individuals would be the first to
respond to task needs; presumably they would
perform proportionately more tasks (the "elites"
described for most social insects [e.g., Oster and
Wilson 1978]). Less sensitive workers make an
effective reserve corps. If active nurses die or
are otherwise removed, brood care needs would go
unmet; the intensity or frequency of brood
requirements would rise above the threshold point
of less sensitive, previously non-nurse workers,
and those workers would then commence brood care.

Variation in age at first performance of a
particular behavior might also reflect such
threshold differences. Alternatively, that
variation may reflect differences in individual
ontogenies. For <u>Solenopsis invicta</u> workers, the
single best predictor of age at onset of any
behavior was age at death (Mirenda and Vinson
1981). Unfortunately, our data do not allow
differentiation between these two possibilities.

Overall age may also affect task rate. Task
rate (frequency of performance per unit time) may
decrease as age increases. All three individuals
performing tasks at below average rates died
relatively old, and #12, with the highest task
rate, died young. That suggests several
explanations. Perhaps the laboratory situation
does not present a full spectrum of task
opportunities for older workers. Or perhaps task
rate alone is an inadequate measure of effective

differences among individuals, and per task
duration should be considered as well. Or
perhaps, in fact, general activity rate and
behavior performance do decline with increasing
age, as has been suggested or implied by others
(Schmid-Hempel and Schmid-Hempel 1984; Porter and
Jorgensen 1981; Oster and Wilson 1978). Again,
although our sample sizes preclude differentiation
among these alternatives, the data certainly
suggest that analyses of activity rate should
include controls for age.

The data also suggest that given ants vary by
month in their proportionate frequencies of
behaviors, and that they do so despite the
(apparent) constancy of colony conditions and
physical caste ratios. Such variation could be
for reasons internal and/or external to the
ants. Workers could show fluctuations in activity
rates or behavior performance for developmental or
ontogenetic reasons, and/or in response to
variations in colony need. If colony conditions
and caste ratios are actually constant, such
variation in need should stem from differences in
whether or how those needs are being met. That
is, the effective rather than actual composition
of workers available to meet those needs
apparently varies; for example, as new workers
eclose into the colony, there is presumably change
among proportions of individuals of various
sensitivities, and thus also change in the
concomitant division of labor.

Finally, let us consider "role." Inspection
of all the data suggests several roles for minor
workers, based on the Oster and Wilson definition
of role as "sets of behaviors linked together by
relatively high transition probabilities"
(1978:122). These roles can have an age
component; that is, workers can progress from one
to another with age. However, some workers
perform more than one role at any given age, while
some never perform particular roles.

Based on data from both known-age and other
workers, among the inside workers there were
nurses (which performed brood care) and non-
nurses, while among outside workers there were
foragers and non-foragers. Thus, being an inside
or outside worker does not itself constitute a

role. But among workers associated with either locale, there are those which perform locale-specific tasks, and those which do not. (Those which do not would, under the different sensitivities model, presumably be individuals with higher thresholds to needs for those locale-specific tasks.)

For the 12 ants, three of five nurses ultimately ceased brood care and foraged, one never foraged, and one performed brood care throughout her life, although she also foraged. This "mix" of inside/outside behaviors was seen also among the workers of unknown ages. Five were excluded from those analyses because they performed tasks in both locales. Were they in "transit" from inside to outside work? That is, would they ultimately have become "pure" outside workers? Need one invoke such transit to explain a mix of inside/outside behavior? One of the known-age individuals (#6) showed such a mix for over 2 of the 3 months of her life. By the transition frequencies definition of role, these workers are performing two roles. Since a worker performing both inside and outside tasks performs clusters of one and then changes locale and performs clusters of the other, the transition frequencies between inside and outside behaviors is less than expected, while those among behaviors within each set are greater than expected.

Thus, one cannot use transition probabilities alone to define roles--apparently one should restrict their use to behaviors performed in particular locales. That has the disadvantage of considerably decreasing the number of behaviors which can possibly be performed, which, in turn, means that role differences will not be seen unless role membership is very clear, and/or sample sizes are astronomical. The latter is particularly true if only task behaviors are considered; since tasks comprise roughly 22 percent of all behaviors, the number of observations must be five times the sample size needed. Further, neither age, nor behavior, nor exclusivity of behavior performance seems tightly linked to these roles.

Based on these data and definitions, we must answer the question, "Is a role the property of an

individual or a caste?" with, "Yes." Apparently
it can be both--or is it therefore neither?

In sum, we find considerable variation in
performance of behaviors and behavioral ontogeny
of known-age individual ants, including whether
they ever perform particular behaviors. Certainly
such variability has far-reaching implications for
assessment of colony labor profile, division of
labor, and productivity.

SUMMARY

Observation of twelve known-age individual
ants of Camponotus sericeiventris Guerin
(Hymenoptera: Formicidae) showed them to be
significantly different behaviorally. We suggest
that some of these differences are due to
"sensitivity" differences among individuals.
Analyses of behavior pair co-occurrences from
those data and data on an additional 68 ants show
that particular pairs co-occur at frequencies not
expected based on the independent frequencies of
the behaviors. Such differences are considered
with respect to "roles."

ACKNOWLEDGMENTS

We thank N. F. Carlin for critical discussion
of and comments on this work, and J. F. A.
Traniello for reading an early draft of the paper.

REFERENCES

Bhatkar, A., and Whitcomb, W.H.   1970.   Artificial
    diet for rearing various species of ants.
    Florida Entomologist 53:229-232.
Busher, C.E., Calabi, P., and Traniello, J.F.A.
    1985.   Polymorphism and division of labor in
    the Neotropical ant Camponotus sericeiventris
    Guerin (Hymenoptera: Formicidae).   Annals of
    the Entomological Society of America 78:221-
    228.

Calabi, P. 1986. Division of labor in the ant
Pheidole dentata: the role of colony
demography and behavioral flexibility. Ph.D.
diss., Department of Biology, Boston
University, Boston, Massachusetts.
Calabi, P. Behavioral flexibility in Hymenoptera:
a re-examination of the concept of caste. In
Advances in Myrmecology, ed. R.H. Arnett,
Jr. Leiden: E.J. Brill. In press.
Calabi, P., Traniello, J.F.A., and Werner, M.H.
1983. Age polyethism: its occurrence in the
ant Pheidole hortensis, and some general
considerations. Psyche 90:395-412.
Herbers, J.M., and Cunningham, M. 1983. Social
organization in Leptothorax longispinosus
Mayr. Animal Behaviour 31:759-771.
Michener, C.D. 1974. The social behavior of the
bees: A comparative study. Cambridge,
Massachusetts: Harvard University Press.
Mirenda, J.T., and Vinson, S.B. 1981. Division
of labour and specification of castes in the
red imported fire ant Solenopsis invicta.
Animal Behaviour 29:410-420.
Oster, G., and Wilson, E.O. 1978. Caste and
ecology in the social insects. Princeton,
New Jersey: Princeton University Press.
Otto, D. 1958. Über die Arbeitsteilung im Staate
von Formica rufa rufo-pratensis minor Gössw.
und ihre Verhaltensphysiologischen
Grundlagen: Ein Beitrag zur Biologie der
roten Waldameise. Wissenschaftliche
Abhandlungen der Deutsche Akademie der
Landwirtschafstwissenschaften, Berlin 30:1-
169.
Porter, S.D., and Jorgensen, C.D. 1981. Foragers
of the harvester ant Pogonomyrmex owyheei--a
disposable caste? Behavioral Ecology and
Sociobiology 9:247-256.
Schmid-Hempel, P., and Schmid-Hempel, R. 1984.
Life duration and turnover of foragers in the
ant Cataglyphis bicolor (Hymenoptera:
Formicidae). Insectes Sociaux 31:345-360.
Wilson, E.C. 1971. The insect societies.
Cambridge, Massachusetts: Harvard University
Press.

Wilson, E.O. 1983. Caste and division of labor
    in léaf-cutter ants (Hymenoptera: Formicidae:
    Atta). III. Ergonomic resiliency in
    foraging by A. cephalotes. Behavioral
    Ecology and Sociobiology 14:47-54.
Wilson, E.O. 1985. The sociogenesis of social
    insect colonies. Science 228:1489-1495.
Zar, J.H. 1974. Biostatistical analysis.
    Englewood Cliffs, New Jersey: Prentice Hall,
    Inc.

Appendix 3.1
Transition frequencies among pairs of all recorded behaviors
for all ants[a]

	Following Act						
Preceding Act[b]	ANT	ELF	ESF	EX	FOR	GRD	GRM
ANT	–	.0001	0	.006	.012	.001	.001
ELF	0	–	0	0	.004	0	0
ESF	0	0	–	0	0	0	0
EX	.002	0	0	–	.006	0	.0009
FOR	.004	.004	0	.006	–	.0002	0
GRD	.0003	0	0	0	0	–	.0001
GRM	.013	0	0	0	.0001	0	–
HNE	0	0	0	0	0	0	0
HNL	0	0	0	0	0	0	.0001
HNN	0	0	0	0	0	0	0
HNP	0	0	0	0	0	0	.0002
LE	.0002	0	0	0	0	0	.0001
LL	.0001	0	0	0	0	0	.0009
LN	0	0	0	0	0	0	.0003
LOC	.008	0	0	.0002	0	.003	.007
LP	.0001	0	0	0	0	0	.0003
PTL	.001	0	0	.02	0	.002	.0005
REG	.05	0	0	.002	.003	.0001	.002
RGL	0	0	0	0	0	0	.0001
RST	.01	0	0	.01	.008	.0001	.009
SG	.005	.002	.0001	.02	.02	.001	.01

(continued)

[a]Based on 10,396 pairs.

[b]See Table 3.1 for a list of behaviors.

Appendix 3.1 (continued)

			Following Act				
Preceding Act	HNE	HNL	HNN	HNP	LE	LL	LN
ANT	0	0	0	0	0	.0003	.0004
ELF	0	0	0	0	0	0	0
ESF	0	0	0	0	0	0	0
EX	0	0	0	0	0	0	0
FOR	0	0	0	0	0	0	0
GRD	0	0	0	0	0	0	0
GRN	0	0	0	0	.0001	.0001	.0001
HNE	–	0	0	0	.002	0	0
HNL	0	–	0	0	0	.0004	0
HNN	0	0	–	0	0	.0002	0
HNP	0	0	0	–	0	0	0
LE	.002	0	0	0	–	.0001	0
LL	0	.001	0	0	.0003	–	0
LN	0	0	0	0	0	0	–
LOC	.0002	.0003	.0006	.0006	.003	.004	.003
LP	0	.0001	0	.0008	0	.007	0
PTL	0	0	.0001	0	0	0	.0001
REG	0	0	0	0	0	0	0
RGL	0	.0001	0	0	0	.005	0
RST	0	0	.0009	.0002	.002	.001	.002
SG	0	0	.001	0	.002	.004	.002

	Following Act						
Preceding Act	LOC	LP	PTL	REG	RGL	RST	SG
ANT	.03	.0002	.004	.001	.0001	.03	.01
ELF	0	0	0	0	0	0	.001
ESF	0	0	0	0	0	0	.0001
EX	.0001	0	.02	.005	0	.01	.02
FOR	0	0	.0001	.01	0	.01	.02
GRD	.005	0	.005	.0004	0	.0007	.002
GRM	.01	.0003	.0001	.002	.0003	.007	.006
HNE	.0003	.0001	0	0	0	0	0
HNL	.001	0	0	0	0	0	.0001
HNN	.0009	0	0	0	0	.0006	.001
HNP	.0007	.0007	0	0	0	0	0
LE	.004	.0001	0	0	0	.0006	.001
LL	.005	.001	0	.002	.003	.001	.002
LN	.004	0	.0001	.0001	0	.001	.002
LOC	–	.005	.006	.02	0	.06	.04
LP	.006	–	0	0	.0003	.001	.003
PTL	.008	0	–	.004	0	.009	.008
REG	.008	0	.001	–	0	.004	.004
RGL	.0008	0	0	0	–	.0004	.0003
RST	.04	.002	.009	.03	0	–	.03
SG	.04	.002	.01	.004	0	.03	–

4

Variation in Foraging Behavior Among Workers of the Ant *Formica schaufussi*: Ecological Correlates of Search Behavior and the Modification of Search Pattern

James F. A. Traniello

Many studies on the organization of insect societies have demonstrated the existence and significance of individual differences in behavior among colony members. Insect societies are strongly compartmentalized, and eusociality, by definition, is characterized in part by patterns of division of labor that reflect the specializations of individuals. Primary patterns of task performance by age- or size-related worker subgroups are well documented in the literature (reviewed by Wilson 1985; Calabi et al. 1983). However, recent research has shown that within a subgroup, workers may show more fine-tuned specializations, resulting in the existence of "specialists among specialists." For example, older workers of many social insect species function as foragers, but within the age group that comprises the forager caste, individuals may differ in their patterns of search behavior or food choice (Heinrich 1976, 1979; Rissing 1981; Wehner et al. 1983; Schmid-Hempel 1984).

How can the significance of individual specialization be analyzed? An approach that places the behavior in question in an ecological framework is obviously necessary. Moreover, because individual specializations most likely result, at least in part, from the different experiences of workers, a perspective incorporating learning would be valuable. In regard to foraging behavior, variation among workers could represent a biologically meaningful component of an individual's foraging strategy.

91

In foraging ecology, a behavioral/ecological
approach to foraging strategy notes that many
environmental details relevant to foraging, such
as food distribution patterns, are largely
unpredictable and must be adjusted to through
experience. Learning, therefore can be viewed as
an adaptation which permits an animal to deal with
environmental unpredictability (Shettleworth
1984). Therefore, studies of individual
variability and learning interface because
different individuals may experience different
environmental details and modify their behavior
accordingly. Through such individual behavioral
adjustments, fitness may be enhanced. In social
insect species in which forager lifespan is
extremely short (Porter and Jorgensen 1981;
Schmid-Hempel and Schmid-Hempel 1984), it would
appear that relatively small differences in
individual experience may result in very different
foraging behaviors.
    In this paper I describe the results of
studies on differences in search behavior among
workers of the ant <u>Formica schaufussi</u> Mayr, and
attempt to show how individual experience and food
resource-related modification of search patterns
may contribute to a worker's search tactics.

STUDY ANIMAL AND GENERAL METHODS

    <u>Formica schaufussi</u> is a common north
temperate open-field ant that nests in soil and
has generalist food habits. Workers tend aphids
for honeydew secretions and feed on plant sap
directly, and also forage for a wide range of
arthropod prey, chiefly dead or dying insects
(Traniello 1980, 1987; Traniello et al. 1984).
Workers actively forage during the cooler portions
of the day in the early morning and late
afternoon, and peaks in foraging activity occur at
0900 and 1700 hours. Seasonally, foraging
activity is at a maximum during late June. All of
my observations and experiments were carried out
from May to September. Field sites were located
at Belmont, Westford, and Natick, Massachusetts.
Workers were marked with Testors® enamel paint to
permit individual recognition. For all

experimental series, each worker was used for only one trial of a given type of test. Statistical analyses were performed using a SPSS program. Pearson correlation coefficients were generated to determine the significance of association between variables. Statistical comparisons were also made using a Mann-Whitney U-test.

VARIATION IN FORAGING BEHAVIOR

Variation in Trip Time, Number, Direction, and Distance

Using individually identifiable color codes, I marked workers and followed their foraging behavior through the morning activity period. Workers depart from the nest entrance in different directions, which appear to be consistent for each individual (at least over short time periods). Each foraging trip is characterized by a travel and search phase. Foragers move straight out from the nest and travel for various distances before beginning to show local search behavior, which is characterized by a high frequency of turning and probing. Travel phase distances (the distance from the nest at which searching first began) and search phase durations to foragers are given in Table 4.1, and the distribution of trip durations and number of trips/forager are given in Table 4.2. These data, which show strong variation among foragers in spatial components of foraging and overall contribution to foraging effort, suggest that colonies may spatially partition search effort, and that workers vary in their foraging "initiative." The variability in "initiative" does not seem to depend upon the success of an individual immediately prior to a given trip. For example, of a total of eight successful foragers that left the nest after returning with prey, five made no second trip on that day.

Table 4.1
A comparison of individual foraging trips of Formica schaufussi at the Westford and Belmont sites[a]

Component	N	Mean	SD	Range[c]
		Westford		
Travel phase distance (m)	22	5.2*	± 5.6	0-24
Travel phase duration (min)	22	6.9[b]	± 6.8	0-22.5
Search phase duration (min)	11	15.1[b]	± 13.1	1.5-45.8
		Belmont		
Travel phase distance (m)	17	1.7*	± 2.1	0-7.6
Travel phase duration (min)	17	4.0[b]	± 5.5	0-17
Search phase duration (min)	6	22.2[b]	± 26.5	3.2-74

[a]Differences between the two sites were analyzed using a Mann-Whitney U-test.

[b] = NS, * = P < .01.

[c]Travel phase distances and durations equal to 0 correspond to cases in which a forager immediately began searching after leaving the nest.

Table 4.2
Frequency distribution of foraging trip
durations (A) and the number ot trips
made (B) by F. schautussi workers at
Westford

A

Trip Duration (min)	No. of Workers	Percent
< 15	16	28
16-30	8	14
31-60	14	25
61-120	14	25
121-240	5	8
Total	57	100

B

No. of Trips per Worker	No of Workers	Percent
1	26	53
2	15	31
3	3	6
4	3	6
5	2	4
Total	49	100

Variation in Search Pattern

In addition to the variability in foraging
behavior described above, foragers also differ in
their tendency to return to and continue to search

at a location where food was found on a prior successful excursion. I could demonstrate such differences by recording the travel and search phase patterns of individually marked workers that had just been given a prey item (a small dead insect) upon their return to the site where prey was offered. Based on marked differences in the time allocated to area-restricted search in the vicinity of the prior food find, persistent and non-persistent foragers could be recognized (see also Schmid-Hempel 1984). Two types of non-persistent foragers were observed. Type a showed no search whatsoever in the area where food was previously found, and type b showed area-restricted searching after first returning to the site, but then would soon leave. For both a and b type non-persistent foragers, the bearing of a forager leaving the site of a prior food find was almost always equivalent to the initial direction chosen by a forager as she left the nest entrance. Persistent foragers show a high degree of area-restricted searching in the vicinity of the prior food find and did not depart from the site of the find at their entry angle. A forager was classified as non-persistent if at any time during the five-minute post-return observation period it left the area of the prior find and moved straight out, maintaining an angle approximately equal to the angle of its initial departure from the nest. Generally, type b non-persistent foragers departed from the site of a prior reward within one or two minutes after returning.

CORRELATION BETWEEN RESOURCE TYPE AND PERSISTENT AND NON-PERSISTENT FORAGING

Formica schaufussi, like many other ant species, is omnivorous. Workers tend aphids for sugar-rich honeydew secretions and forage for protein, chiefly in the form of dead insects. These types of food differ considerably in their distribution patterns: aphids are essentially fixed in their spatial location, and their carbohydrate secretions are consistent in quality and are renewed at predictable rates (Mittler

1958; Sudd and Sudd 1985). Insect prey, in
contrast, are unpredictable in time, space, and
size, and generally are not renewed with any
certainty. Given these differences in the
distribution patterns of the food types taken by
F. schaufussi workers, I next examined whether
there was a correlation between the search pattern
of a forager and the type of food collected. In
other words, do persistent and non-persistent
search patterns correlate with the persistence and
non-persistence of a resource?

To answer this question I offered a single 3-
mg termite (a Reticulitermes flavipes [Kollar]
worker) or a 2 µl droplet of 1 M sucrose to marked
workers that traveled 1 m from the nest. Using
forceps, the freshly killed termite was placed in
front of a worker. Sucrose solution was offered
using a syringe, and workers were allowed to feed
on a droplet until repletion (87.1 ± 5.8 sec, mean
± standard error, N = 95). After a worker
returned homeward, the soil surface in the
vicinity of the food offering was scraped clean.
This eliminated any search biases due to the
presence of food odors or chemicals deposited by
foragers. In regard to the latter possibility,
workers trail-laid in only 5.3 percent of
occasions (N = 95) in response to 1 M sucrose and
8.8 percent of occasions (N = 113) after receiving
a termite. It is, therefore, unlikely that the
results were biased by the deposition of a
chemical cue that might induce search behavior.
Honing time, time spent in the nest, return time,
and temperature were recorded. Differences in
persistence were quantified during the first five
minutes after a worker returned to the site of a
prior food find by recording a number of variables
that relate to search pattern. Search persistence
was measured as the amount of time spent searching
in a circular area of radius 20 cm having the
locus of the food find as its center. Components
of foraging behavior measured after a worker
received either a termite or a sucrose feeding
were:

1. Initial Giving up Time (IGT): the time
   spent searching in the circular area after
   first returning to within 10 cm of the

location of the prior food find to the
first departure outside of the circular
area; not measured for type a non-
persistent foragers; measured for type b
non-persistent and persistent foragers.

2. Area-Restricted Searching (ARS): the time
spent searching within the circular area
in the vicinity of a prior find. The
phenomenon is consistent with that
described by Curio (1976), Bond (1980),
Bell (1985), Bell et al. (1985), and
Jander (1975).

3. Sector-Restricted Search (SRS): the time
spent searching in a 45° sector about the
location of a prior find (see also
Harkness and Magroudas 1985).

4. Out-of-Sector Search (OSS): the time
spent searching outside of the above-
mentioned 45° sector.

5. Maximum Distance from Site of Prior Find
(Dmax): the maximum distance away from
the site of the prior food find; measured
after marking a forager's path with color-
coded toothpicks.

6. Maximum Angular Deviation (∠max): the
angle between Dmax and the site of the
prior find.

The results obtained indicate that workers
showed different degrees of persistence in
response to collecting each type of food source,
and these differences were not correlated with
temperature, contrary to what one might have
expected in a species in which temperature has
been shown to have an effect on foraging and prey
selection (Traniello et al. 1984, 1988). Only
the maximum angle of deviation was correlated with
temperature, but this one component of search was
not related to food type (Table 4.3; later refer
to Tables 4.5 and 4.6 for verification). The
differences in the most readily measured and
meaningful aspect of search persistence--area-
restricted search--in response to the two food
types are interesting. First, persistent and non-
persistent foraging patterns occur in response to
both protein and carbohydrate food, and for each

Table 4.3
Pearson correlation coefficients between
temperature and components of search behavior

Component of Search Behavior	N	r	P
Persistence or non-persistence	80	.08	.235
Initial giving up time	59	-.13	.160
Area-restricted search	69	-.08	.253
Sector-restricted search	41	.19	.113
Out-of-sector search	46	-.04	.403
Dmax	33	.06	.366
Lmax	34	-.36	.020

food type persistent foragers obviously spend
significantly more time in area-restricted search
than non-persistent foragers (Table 4.4). These
data quantitatively substantiate the behavioral
categorization of forager types. Most importantly
in terms of the effect of resource quality on
search persistence, persistent foragers that had
fed on surcrose showed a greater area-restricted
search effort than those fed a termite, and even
non-persistent foragers (type b) spent more time
searching before their departure if offered
surcrose. The former difference is statistically
significant (P = .027), the latter approached
significance (P = .068).
    Comparison of other components of foraging
used to quantify search persistence also revealed
significant resource-correlated differences in
foraging behavior. Initial giving up time was
significantly greater in response to sucrose for
both persistent and non-persistent foragers, and
area-restricted search and sector-restricted
search were also significantly greater in response
to sucrose (Table 4.5). Time spent in out-of-
sector search was significantly lower, and the
maximum distance from the site of a prior find was
significantly shorter when workers fed on

Table 4.4
Area-restricted search effort (mean ± SE, in min)
for persistent (+) and non-persistent (-) foragers
feeding on sucrose or termites[a]

Forager Type	$\underline{N}$	Mean (min)	SE
		Sucrose	
Persistent	21	$218.0^{b,d}$	± 19.8
Non-persistent	17	$64.2^{b,e}$	± 14.7
		Termite	
Persistent	23	$151.7^{c,d}$	± 17.4
Non-persistent	23	$17.6^{c,e}$	± 5.9

[a]Statistical comparison by a Mann-Whitney U-test.

[b]$\underline{P} < .0001$

[c]$\underline{P} < .0001$

[d]$\underline{P} = .027$

[e]$\underline{P} = .068$

sucrose. The time spent in sector-restricted
search and out-of-sector search reflect a phase of
relatively straight locomotion equivalent to the
ranging behavior described by Jander (1975).
Maximum angular deviation was slightly lower in
response to sucrose, but this difference was not
significant (Table 4.5).
    Approximately equal numbers of workers showed
persistent and non-persistent behavior in response
to carbohydrate and protein food. The differences
in food type, however, produced differences in the
degree of persistence shown by workers that have

Table 4.5
Resource-related differences between foragers in search behavior.

Foraging Component[a]	Food Type						P[b]
	Termite			Sucrose			
	N	Mean	SE	N	Mean	SE	
IGT(−) (min)	12	30.4	± 3.8	12	75.8	± 15.8	.02
IGT(+) (min)	21	109.9	± 22.5	21	178.1	± 26.5	.04
ARS (min)	23	151.7	± 17.4	21	218.8	± 19.8	.03
SRS (min)	23	262.9	± 12.8	21	290.8	± 5.1	.12
OSS (min)	23	35.1	± 11.5	22	8.8	± 4.8	.05
Dmax (cm)	23	62.0	± 5.4	21	41.9	± 5.4	.01
∠max (°)	21	23.9	± 3.9	21	18.8	± 3.3	.31

[a] See text for a description of foraging components. (+) = persistent foragers; (−) = non-persistent foragers.

[b] Statistical comparison by a Mann-Whitney U-test.

contacted food of each type. There were
significant differences in the number of workers
showing total, 75 percent, and 50 percent
persistence in response to each food type (Table
4.6). Whereas 9.3 percent of workers did not
return to the site of a prior food find when a
termite was offered. These results indicate that
foragers show a greater degree of search effort in
response to the food type which should have a
greater payoff if greater effort is allocated.
That is, search persistence is greater in response
to feeding on a persistent resource.

RESOURCE-RELATED MODIFICATION OF LOCAL SEARCH

I next questioned if the development of
persistent and non-persistent foraging
specializations were influenced by food type. In
other words, are workers predisposed to modifying
their search behavior according to the type of
food encountered? Such predispositions are known
to occur in other insects (Menzel 1984). I tested
this possibility in the following way.
After a single reward of either protein or
carbohydrate food, directional- and distance-
sensitive components of a worker's search pattern
were quantified as described above. This allowed
workers to be categorized as either persistent or
non-persistent foragers, and their search effort
quantified. After the five-minute period elapsed
during which search effort was recorded, workers
were given three successive rewards (either three
termites or three sucrose feedings) without any
intervening measurement of foraging pattern.
After the third reward, a worker's behavior was
recorded upon return to the reward site, and again
their search effort was categorized and
quantified. Although sample sizes were small,
some interesting differences and trends were
evident when protein- and carbohydrate-rewarded
foragers were compared, suggesting that some
components of search behavior were more sensitive
to modification than others. Initial giving-up
time and maximum distance from the site of a prior
find, both of which were significantly different
after a single offering of protein or

Table 4.6
Relationship of search persistence to food type

Foraging Behavior	Food Type					
	Termite		Sucrose			
	N	Percent	N	Percent	p^b	
Persistent	23	39.7	21	48.8	NS
Non-persistent	23	39.7	18	41.8	NS
Return to site	12	20.7	4	9.3	NS
Totally persistent[a]	2	3.4	9	20.9	.004
75% persistent[a]	4	6.9	10	23.3	.017
50% persistent[a]	10	17.2	16	37.2	.023
Total	58		43		

[a]percent persistence is calculated as the fraction of the 5-minute period during which area-restricted search behavior occurred.

[b]Statistical comparison by a Mann-Whitney U-test.

carbohydrate, were not significantly different after three offerings, although responses to each food type showed a trend in the right direction (Table 4.7). Area-restricted search and sector-restricted search were longer in duration in response to sucrose, and out-of-sector search was shorter. However, these differences only approached significance at the .05 level. Maximum angular deviation was about equal. Twenty percent of workers shifted from non-persistent to persistent foraging when given protein (N = 20) whereas 44 percent of workers showed such a change when given sucrose (N = 23). Also, 5 percent of workers shifted from non-persistent search to total persistence when given protein; 18 percent shifted in response to sucrose. Finally, 15 percent of non-persistent workers offered termite rewards remained non-persistent; no workers given sucrose remained non-persistent. Significantly, more workers showed at least 75 percent persistence in foraging and spent a greater total time in sector-restricted search after three successive sucrose rewards (Table 4.8).

DISCUSSION

Workers of Formica schaufussi differ in the extent to which they will continue to search in the vicinity of a site where food was previously found. The type of food encountered strongly influenced the tendency of a worker to search the site of the find, but foragers given one reward of either protein or carbohydrate food showed persistent or non-persistent search patterns independent of reward type. Although the behavior of type b non-persistent foragers might appear to represent a transitional stage during a shift from non-persistent to persistent search, there is currently no information on whether such a progression of change in foraging strategy actually occurs. Variability in search pattern among workers may represent an adaptation to efficiently harvest two types of food with very different temporal and spatial distribution patterns. It is reasonable to assume that the workers responding to a persistent food type by

Table 4.7
Resource-related modification of foraging behavior

Foraging Component[a]	Food Type						
	Termite			Sucrose			
	N	Mean	SE	N	Mean	SE	p[b]
ICT (min)	17	75.1	± 16.2	19	121.3	± 23.2	.381
ARS (min)	17	132.8	± 19.4	18	205.3	± 20.9	.068
SRS (min)	17	275.8	± 8.2	18	294.7	± 3.6	.066
OSS (min)	17	24.2	± 8.2	17	5.6	± 3.8	.066
Dmax (cm)	16	62.8	± 5.6	16	54.1	± 6.9	.352
∠max (°)	16	14.2	± 3.5	18	13.3	± 2.7	.420

[a] Search pattern was recorded after three successive rewards of a given type. See text for description of foraging components measured.

[b] Statistical comparison by a Mann-Whitney U-test.

Table 4.8
Resource-related differences between workers in area-restricted search patterns

Foraging Behavior	Food Type					
	Termite (N = 17[a])		Sucrose (N = 18[a])			
	No. Responses	Percent	No. Responses	Percent	p	
75% persistence	3	17.6	11	61.1	.006	
50% persistence	7	41.2	13	72.2	.060	
Total time spent in SRS	9	52.9	16	88.9	.014	

[a]N = total number of individuals tested.

returning to the site of the find and continuing
to search the area will have a higher payoff in
foraging success than those that show a
temporarily limited search. Likewise, foragers
that have found insect prey will have little
success in locating additional prey of the same
type in the same area because of the uncertainty
of dead insects in time and space. Also, workers
show a tendency to return to a site where
persistent, but not non-persistent, food was
found. However, the major difference between non-
persistent and persistent foragers is not only
their tendency to return to and search the site of
a previous food find, but also the degree to which
they allocate search time and the spatial accuracy
of their search. Similar search patterns have
been described in the ant Cataglyphis bicolor
Fabricius (Schmid-Hempel 1984), in which search
behavior was affected by the type of food
contacted. However, the response of individual
workers was variable. Because one of the food
types (a piece of cheese) used in Schmid Hempel's
experiments was "evolutionarily unfamilar" to
these desert ants, it is difficult to determine
how an individual should adjust its search pattern
after locating such an item. However, all
foragers persisted less in re-searching the site
of a find of a non-persistent food item (a fly).
The profitability of the items offered may also
have influenced search behavior.

F. schaufussi workers recognize protein and
carbohydrate food and seem to adjust their search
behavior in a fashion that could increase
energetic efficiency. Moreover, the modification
of an individual's local search pattern may be
facilitated by reinforcement with a reward of a
food with a spatially and temporally predictable
distribution. However, experience during callow
development may also influence the search pattern
of mature workers, and this possibility cannot be
ruled out. Although individual workers showed an
increase in the degree of persistence of foraging
when they received consecutive rewards of either
sucrose or a termite, the amount of time spent in
area-restricted or sector restricted search was
greater and out-of-sector search was shorter when
workers were rewarded with sucrose. Other

components of search behavior, such as initial
giving-up time and the spatial sensitivity of the
search as reflected in the forager's maximum
distance from the reward site do not appear to be
as sensitive to the type of reward used as
reinforcement. Most suggestive, however, is the
fact that more workers shifted their strategy from
non-persistent to totally persistent search and
fewer workers that were initially categorized as
non-persistent remained so when sucrose was used
to reinforce a forager's return to a site.
Rewards of insect prey were not as effective in
causing workers to alter their search behavior.
It appears that workers may be predisposed toward
developing search persistence depending upon the
type and frequency of food contacted.

Does the modification of search behavior
described in these studies reflect only an
ephemeral change in local search or suggest
learning? I have recently been able to show that
search accuracy does not increase as a function of
repeated reinforcement with either protein or
carbohydrate food. However, carbohydrate food
elicits a significantly more accurate search.
Reinforcement with insect prey produces a highly
variable, less spatially accurate change
characterized by greater ranging during search.
Also, time spent in area-restricted search is
highly variable in response to rewards of insect
prey and workers show little or no increase in ARS
in response to multiple rewards. A reinforcement
schedule of 0.25, 0.5 or 1 M sucrose, however,
produces a more consistent, higher level of ARS.

Are there age and size correlates to forager
search patterns in F. schaufussi? Differences in
body size among foragers are not pronounced, but
the possibility that slight variation in body size
may be associated with a given search persistence
predisposition remains to be tested. Also, search
behavior may be age-dependent. Schmid-Hempel
(1984) considered the six-day lifespan of
Cataglyphis bicolor foragers too brief to account
for differences in foraging strategy among workers
in this species. However, F. schaufussi foragers
have a longer lifespan (approximately 20 days).
If foragers showing different search patterns are
differentially prone to predation risk, the

strategy with the greater associated risk may be
scheduled later in a forager's life to maximize
its labor contribution to the colony. It will be
important to document whether a given search
tactic categorizes an individual throughout its
tenure as a forager, thereby reflecting search
specialization.

While the initial data set presented in this
paper provides evidence supporting the hypothesis
that foraging behavior may result from
ecologically correlated behavioral predispositions
and resource-related experiential processes, the
relationship of search persistence and resource
persistence requires further consideration. For
example, large insect prey should not have a
greater reinforcement value in the development of
search persistence because insect prey,
independent of their size, are unpredictable in
time and space. Also, memory duration should be
related to the type of reward used to reinforce
search persistence. Information on the spatial
distribution of persistent food such as aphids
should be retained for longer periods of time than
information on the location of insect prey, even
if the prey are spatially clumped. Such prey,
like termite alates, are patchily distributed and,
once exploited, are not renewed with any
certainty. It may be that learning rate is not
different for a given number of carbohydrate or
protein rewards because search persistence should
be easily modified, at least in the short-term, to
efficiently exploit clumps of either food type.
Memory duration, however, should be shorter in
response to reinforcement of search persistence
with insect prey: clumps of non-persistent food
are themselves non-persistent. Food distribution
pattern, worker lifespan, the differential
reinforcement value of protein and carbohydrate
food, memory duration and predispositions in
foraging specialization may all be associated, but
the relationships among these behavioral and
ecological factors remain to be examined. In
particular, the relationship between local search
and learning and its ecological correlates require
considerable further attention.

SUMMARY

Workers of the ant Formica schautussi show variation in several components of foraging behavior. Foragers leave the nest in ditferent directions and travel up to 40 m before beginning area-restricted search. The number and duration of foraging trips also varies among workers. Foragers differ in their tendency to return to a site where food had been previously found and continue to search there. Based on the strong differences in the duration of area-restricted search in the vicinity of a prior find, foragers could be categorized as persistent or non-persistent. Non-persistent foragers showed no or little search at the location when food was found, whereas persistent foragers intensively searched the area. An approximately equal number of foragers showed each type of search pattern.

Search duration was correlated with the type of food that a forager collected. Persistent foragers searched significantly longer it fed sucrose than insect prey, and some components of search behavior (initial giving-up time, out-of-sector search, and maximum distance from find site) were also associated with food type. The local search pattern of a forager could be modified by successive rewards of either carbohydrate or protein food, and individuals rewarded with sucrose tended to increase the duration of their area-restricted search. Forager search patterns appear to be adjusted to the predictability of a resource, and the degree of modification of a forager's pattern seems to depend upon food type.

ACKNOWLEDGMENT

I thank Betty Ferster and Rhys Bowen for assisting in data collection, and Elissa Landre of the Massachusetts Audubon Society's Broadmoor Wildlife Sanctuary for use of field sites. Michelle Scott kindly provided criticism and comments on the manuscript. This research was supported by NSF Grant BNS 82-16734.

REFERENCES

Bell, W.J. 1985. Sources of information controlling motor patterns in arthropod local search orientation. Journal of Insect Physiology 31:837-847.

Bell, W.J., Tortorici, C., Roggero, R.J., Kipp, L.R., and Tobin, T.R. 1985. Sucrose-stimulated searching behaviour of Drosophila melanogaster in a uniform habitat: modulation by period of deprivation. Animal Behaviour 33:436-448.

Bond, A.B. 1980. Optimal foraging in a uniform habitat: the search mechanism of the green lacewing. Animal Behaviour 28:10-19.

Calabi, P., Traniello, J.F.A., and Werner, M.H. 1983. Age polyethism: its occurrence in the ant Pheidole hortensis and some general considerations. Psyche 85:395-412.

Curio, E. 1976. The ethology of predation. Berlin: Springer-Verlag.

Harkness, R.D., and Maroudas, N.G. 1985. Central place foraging by an ant (Cataglyphis bicolor Fabr.): a model of searching. Animal Behaviour 33:916-928.

Heinrich, B. 1976. Foraging specializations of individual bumble bees. Ecological Monographs 46:105-128.

Heinrich, B. 1979. "Majoring" and "minoring" by foraging bumblebees Bombus vagans: an experimental analysis. Ecology 60:245-255.

Jander, R.J. 1975. Ecological aspects of spatial orientation. Annual Review of Ecology and Systematics 6:171-188.

Menzel, K. 1984. Learning in honey bees in a behavioral and ecological context. In Experimental behavioral ecology and sociobiology, ed. B. Hölldobler and M. Lindauer, 55-74. Sunderland, Massachusetts: Sinauer.

Mittler, T.E. 1958. Studies on the feeding and nutrition of Tuberolachus salignus (Gmelin) (Homoptera: Aphididae). III. The nitrogen economy. Journal of Experimental Biology 35:626-638.

Porter, S.D., and Jorgenson, C.D. 1981. Foragers of the harvester ant Pogonomyrmex owyheei: a

112

disposable caste? Behavioral Ecology and
Sociobiology 9:247-256.
Rissing, S.W. 1981. Foraging specializations in
individual seed-harvesting ants. Behavioral
Ecology and Sociobiology 9:149-152.
Schmid-Hempel, P. 1984. Individually different
foraging methods in the desert ant
Cataglyphis bicolor (Hymenoptera:
Formicidae). Behavioral Ecology and
Sociobiology 19:263-271.
Schmid-Hempel, P., and Schmid-Hempel, R. 1984.
Life duration and turnover of foragers in the
ant Cataglyphis bicolor. Insectes Sociaux
31:345-360.
Shettleworth, S.J. 1984. Learning and behavioral
ecology. In: Behavioural ecology: An
evolutionary approach, ed. J.R. Krebs and
N.B. Davies, 170-196. Sunderland,
Massachusetts: Sinauer.
Sudd, J.H., and Sudd, M.E. 1985. Seasonal
changes in the response of wood ants (Formica
lugubris) to sucrose baits. Ecological
Entomology 10:89-97.
Traniello, J.F.A. 1980. Studies on the
behavioral ecology of north temperate ants.
Ph.D. diss., Harvard University, Cambridge,
Massachusetts.
Traniello, J.F.A. 1987. Comparative foraging
ecology of north temperate ants: the role of
worker size and comparative foraging in prey
selection. Insectes Sociaux 34:118-130.
Traniello, J.F.A., Fujita, M.S., and Bowen, R.V.
1984. Ant foraging behavior: ambient
temperature affects prey selection.
Behavioral Ecology and Sociobiology 15:65-68.
Traniello, J.F.A., Ferster, B., and Fujita, M.S.
1988. Thermal correlates of foraging
behavior in the ant Formica schaufussi.
Oecologia. Submitted.
Wehner, R., Harkness, R., and Schmid-Hempel, P.
1983. Foraging strategies in individual
searching ants Cataglyphis bicolor. The
Hague: W. Junk Publishers.
Wilson, E.O. 1985. The sociogenesis of insect
colonies. Science 228:1489-1495.

5

Individual Differences in Social Insect Behavior: Movement and Space Use in *Leptothorax allardycei*

Blaine Cole

The dominance of the superorganism concept for many years contributed to the lack of search for individual differences in the behavior of social insects. With the emergence of a completely new generation of investigators trained in the tradition of kin selection theory and its emphasis on the behavior of individuals, more attention has been paid to the behavior of individuals with the result that differences have been discovered.

The conclusion of individual differences in behavior can be made for at least the following four reasons:

1. Inappropriate comparison
2. Statistical variability
3. Colony-level modification of behavior
4. Individual fitness differences

The first category involves differences that result from misapplication of comparison. The second category involves that phenotypic variability which does not have a heritable genetic component and which, thus, does not respond to natural selection. The latter two categories of individual differences are the result of selective changes in behavior that are the result of selection operating primarily at the between-colony level (category three) or the within-colony level (category four).

113

1. Inappropriate Comparison

If one observes that different individuals
differ in behavior it may be due to something as
overtly incorrect as comparing behavior of one
individual at night and another individual during
the day. Comparisons must be made between
comparable categories of individuals. For
example, to note that there are differences in
behavior between the major and minor worker caste
of an ant is an important observation, but
probably does not say anything useful or
interesting about individual differences in
behavior.

Inappropriate comparisons can be much more
subtle. Suppose there is an ontogenetic
trajectory for a particular behavior, such as
shown in Figure 5.1A. An observation of
individual differences in the behavior of a
younger ant and an older ant may simply be due to
differences in age. They may have identical
behavior if it is integrated over the lifetime.

In order to demonstrate individual
differences in behavior one must be able to
demonstrate that the differences remain when the
entire ontogeny is considered. This can be
difficult, especially when dealing with
hierarchical differences in behavior. It seems
obvious that so clear a difference in behavior as
that between the highest ranking individual,
alpha, and lower ranking individuals is a
straightforward manifestation of individual
differences in behavior. However, if hierarchy
rank is perfectly age-graded, and all individuals
progress up the hierarchy until they become alpha,
then this is simply a manifestation of ontogenetic
stage rather than fundamental differences in
individual behavior.

2. Statistical Variability

It is highly unlikely that ontogenetic
trajectories would be the same for a number of
individuals even if all individuals have the
identical capacity to express the trait (Figure
5.1B). Subtle differences in the social

environment of each individual which probably
influence behavioral development will produce a
frequency distribution for any given behavioral
character. In a large sample of individuals the
difference between the most extreme individuals
may be quite large.

Such individual differences in behavior may
have functional consequences for the colony but
may not themselves be subject to further
analysis. It is important to distinguish this
sort of individual variation in behavior from the
case in which the variability in behavior is
itself a character under colony-level control.
Under some circumstances a highly plastic behavior
may be advantageous to the colony; there could be
colony-level selection on the variability per se
(see case 3 below). However, individual
differences in behavior will exist even in the
absence of this effect.

3.   Colony-level Modification of Behavior

If, in response to a nutritional, pheromonal,
or other colony-level stimulus, the ontogeny of a
behavior in various individuals is changed then
there will be individual differences in behavior
(Figure 5.1C). These differences may be subject
to colony level selection in the same way that
caste ratios may be modified to increase overall
colony efficiency. This phenomenon can be
distinguished from statistical variability since
there is a discrete stimulus or suite of stimuli
that produce a particular behavioral phenotype.
Genetic changes may occur in the extent to which a
character can be modified by a colony-level
stimulus. There may be selection either to
increase or to decrease the malleability of the
response.

Selection is operating at the between-colony
level. Selection is operating to improve colony
functioning and colony efficiency. Those colonies
that have higher foraging efficiency, efficiency
at nest construction or nest defense, etc. will
leave more offspring relative to other colonies.

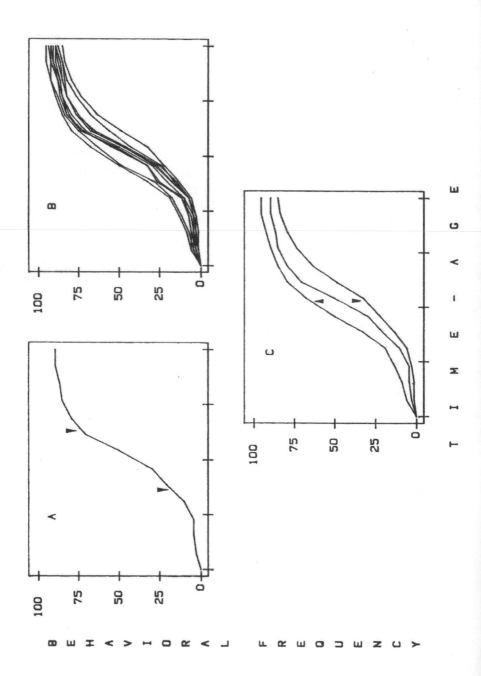

Figure 5.1. The relationship between the frequency of a hypothetical category of behavior (y-axis) and the age of the social insect. A. The ontogenetic trajectories may be the same for two individual workers, but if they are assayed at different ages, they will perform an activity at different rates. B. No two individuals will have the same behavioral ontogenetic trajectory. There will inevitably be statistical, purely phenotypic variability with no heritable component. C. Colony-level selection can modify a typical ontogenetic trajectory in response to specific colony-wide cues.

4.   Individual Fitness Differences

Individual differences in behavior can result from competition among individual social insects for reproductive success. If the colony is limited in its reproductive output and if there is variation in reproductive success among individuals, there will be selection for characters that increase individual fitness, even at the expense of colony reproductive success (Cole 1986). The inevitable result will be individual differences in behavior with the outcome of the competition being expressed as agonistic behavior. I shall describe a case of this fourth type of individual differences in social insect behavior in this paper.

The perception of ant workers as self-sacrificing automatons had a long, distinguished history. Individual workers were considered to be units of the soma of the colony superorganism (Wheeler 1928). Sacrifice of workers was permitted as long as the colony, as an organism, could continue to survive. That perception was dramatically deposed with the advent of kin selection theory. The dominant perception in recent years has been one in which ant workers are maximizing their inclusive fitness by adjustment of foraging strategy, caste ratios, and sex ratio.

One extreme of this viewpoint essentially reclaims the conclusions of the superorganism concept. When the queen is the only reproductive, the workers may benefit by foraging to exhaustion, self-starvation, and suicidal attacks on nest invaders. The critical difference is that workers will be completely self-sacrificing when it is to their advantage to do so. Worker fitness is maximized when colony fitness is maximized, even if it results in the death of a worker.

When workers reproduce there exists the possibility for a tradeoff between selection at the between-colony level and the within-colony level. Although reproduction by workers is an explicit possibility in kin selection theory, there has been little attempt to explore the consequences of worker reproduction for colony function. In the first place there is little known about the nature of any tradeoff between

selection at individual and colony levels. In the
second place there is nothing known about the
consequences of the conflicts of interest due to
reproductive competition between workers in an
insect society.

In this paper I shall begin to examine some
of the consequences of the tradeoff between
reproductive success acquired through personal
reproduction and through helping a close relative
to reproduce. In the ant Leptothorax allardycei
(Mann) workers both reproduce on their own and
help the queen to raise offspring. The present
paper concerns the use of space and the movement
patterns of workers within colonies of this
species. I shall argue below and elsewhere (Cole
1986) that the workers are behaving in such a way
as to maximize their own personal reproductive
success even to the extent of reducing the
reproductive success of other members of the
colony.

The distribution of L. allardycei is in
Southern Florida and the Florida Keys, the
Bahamas, and Bimini. What follows is information
obtained from colonies collected in the Florida
Keys, some of which has been reported in Cole
(1981). Within its geographic range, L.
allardycei is found most commonly in pine-palmetto
scrub forest, where it is a moderately common ant
within its preferred microhabitat. Leptothorax
allardycei places its nests most frequently within
the hollow internodes of sawgrass (Cladium
jamaicense Crantz) culms.

Aggressive interactions between workers are
organized into a dominance hierarchy. An
aggressive interaction between ants usually begins
with antennal contact by the dominant, which then
pummels the subordinate with its antennae and jabs
it with its mandibles. Occasionally the dominant
ant bites the antennae, head, or legs of the
subordinate. The subordinate remains motionless
during the encounter, crouching to the nest
surface and moving only after the dominant worker
moves away.

High dominance hierarchy rank correlates with
high ovarian development (Cole 1981). High
dominance rank is also correlated with the
direction of liquid food flow in the colony.

Liquid food is passed from ants of lower to ants of higher rank. This corresponds to similar observations made on food flow in wasps (Jeanne 1972; Strassmann 1981a, 1981b; West 1967; West Eberhard 1969, 1977).

MATERIALS AND METHODS

Colonies of <u>Leptothorax allardycei</u> were collected from Sugarloaf Key, Florida, in the Florida Keys. The colonies were transferred from the sawgrass stems, where originally found, into glass observation nests. The observation nests were constructed of two glass microscope slides held apart by a 1.5 mm cardboard partition. Nests were housed in plastic boxes with food and water sources. Colonies were fed honey water as well as pieces of insects. The nest was divided into quadrats approximately 2.5 x 4 mm for purposes of recording position.
Individuals of two naturally occurring queenless colonies were marked individually with droplets of paint (Cole 1981). Observations were made of individual ants for periods of one hour. The movements of an ant were continuously recorded together with any activity of the ant and the identity of any ants with which an interaction took place.
During periods of quiescence the position of each identifiable ant was recorded, as were the quadrats occupied by eggs, larvae, and pupae. To compute distance between individuals the center to center distance of each quadrat was used. The average distance between ants in quadrats is the same as the distance between the centers of the quadrats if the positions occur with equal frequency over the quadrat. However, when two objects were within the same quadrat the distance was taken to be zero.

RESULTS

In order to understand the patterns of movement and space use one needs to take into account the aggressive interactions between

individuals and where these interactions occur. A
certain fraction of the interactions between
individuals are of high intensity, involving
biting of some part of the subordinate's body.
Alpha is significantly more likely to be involved
in high interactions with beta (11/134 = 0.082)
than with gamma (3/197 = 0.015; percentage test
t = 2.99, $\underline{P}$ < .01).

When the ants are near the eggs, dominance
interactions take place more frequently than is
expected. The expected number of interactions in
close proximity to the eggs is calculated as
follows. The probability with which a focal ant
is found within a square with eggs, and an object
ant is no further than one square from the focal
ant, is estimated as the sum of the joint
probabilities of occupying these squares (from the
occurrence probabilities below). The expected
frequency of dominance encounters in the vicinity
of the eggs is the product of the total number of
dominance interactions of the individual and the
probability of joint occurrence. Table 5.1 shows
the observed and expected number of dominance
interactions taking place in the vicinity of the
egg squares. Note that the association between
dominance and egg squares may result either if
pairs of ants move non-independently (which would
alter the estimated frequency of joint occurrence)
or if movement was independent and agonistic
levels increased near the eggs.

Movement

The number of times that an ant occupied a
given square within the nest is given in Figure
5.2. I term this occurrence frequency. The
Spearman rank correlation between occurrence
frequency of each ant and the occurrence frequency
of eggs is shown in Table 5.2. Four of the six
ants, and especially alpha, show a significant
positive association between the position of the
eggs and the frequency of occurrence in a given
square. While occurrence frequency is correlated
with egg position, one cannot be used to predict
the other (chi-square for all pairwise comparisons
significantly different, $\underline{P}$ < .05).

2	21	51	37
3	36	92	75
6	21	55	67
2	8	25	27

α - 798

4	11	15	21
9	31	31	39
13	24	29	31
13	8	10	9

β - 798

0	12	29	29
0	20	32	34
6	24	27	31
5	11	18	19

γ - 798

1	18	39	25
19	40	66	41
15	10	10	9
6	7	4	4

α - 795

12	21	12	8
19	27	33	10
20	3	25	18
16	9	13	14

β - 795

19	31	22	23
32	56	65	36
44	31	39	27
30	29	19	15

γ - 795

Figure 5.2. Occurrence frequencies for the three highest-ranking worker ants of two colonies (795 and 798). The number refers to the number of times that a given individual moved into a given nest square. The area shown is the half of the nest that was almost exclusively used by the ants (approximate dimensions 10 x 16 mm.)

Table 5.1
Observed and expected (in parentheses) number of
dominance encounters that occur in proximity to
the eggs

Colony 798			Colony 795		
	β	γ		β	γ
α	17 (8.0)	51 (11.4)	α	20 (2.8)	19 (5.2)
β	-	3 (2.2)	β	-	3 (2.2)

At a more dynamical level one can examine the
transition probabilities from one square to the
next. In the following discussion I will look at
both the one-step (for example, square A to square
B) and the two-step (for example, square A to B to
C) transition probabilities. If the transition
probability between two squares is independent of
the previous location of the ant, movement is
defined as Markovian. That is, movement from
square A to B is independent of whether the ant
was previously in square B or square C. One set
of transitions may be still much more frequent
depending upon the relative frequencies of the
starting squares. Here I emphasize that a
Markovian movement pattern is only one of an
infinity of possible movement rules. However, it
represents a null hypothesis of complete
independence of movement or lack of memory in
movement.
     If one assumes a Markovian movement pattern,
one can calculate the frequency of the two-step
movement patterns based upon the frequency of the
one-step movement patterns and the relative
frequency of occurrence within each given
square. The expected frequency for 500 two-step
transitions was calculated and compared with the
observed two-step transition frequencies. If the
Markov model of movement is correct, then the

Table 5.2
Rank correlation between occurrence frequency of
ant and eggs

Colony	Ant	Correlation
798	α	0.729
	β	0.404
	γ	0.603
795	α	0.828
	β	-0.042 NS
	γ	0.294 NS

distribution of residuals should be normal with a
mean of zero.

Figure 5.3 shows the distribution of
residuals. The mean of the distribution (-0.04)
is near zero with no significant skewness (y_1 =
0.094, t = 0.85, $\underline{P}$ = .4). However the
distribution of residuals is strongly leptokurtic
(y_2 = 2.84, t = 12.85, $\underline{P}$ < .001). A leptokurtic
distribution indicates too many residuals near the
mean and too many residuals in the tails.

Large positive residuals indicate transitions
occurring far more frequently than expected and
may indicate areas of attraction to an ant within
the nest. I chose +2.5 to define a large residual
and asked which transitions these aberrant values
indicated. Figure 5.4 shows the zones of
attraction for alpha and for beta.

The zones of attraction for the alpha and
beta workers seem to differ; for alpha they are in
the vicinity of the eggs and for the beta worker
they seem to be away from the eggs. However, most
of the movements of alpha and fewer of the
movements of beta are found near the eggs, so it
is possible that this is merely a random subset of
all the transitions. To test this, the overall
frequency of occurrence of the alpha workers
within egg-squares (eggs present more than 25
percent of the time) was compared with the
frequency with which the aberrant two-step
transitions occurred in egg squares by a

percentage test. While the alpha workers showed a significantly greater fraction of aberrant two-step transitions occurring within egg-squares (391/833 = 0.469 is the overall frequency of occurrence in egg squares versus 14/18 = 0.778 for the frequency of aberrant two-step transitions within egg squares t = 2.73, P < .01), the beta workers showed a significantly lower fraction of

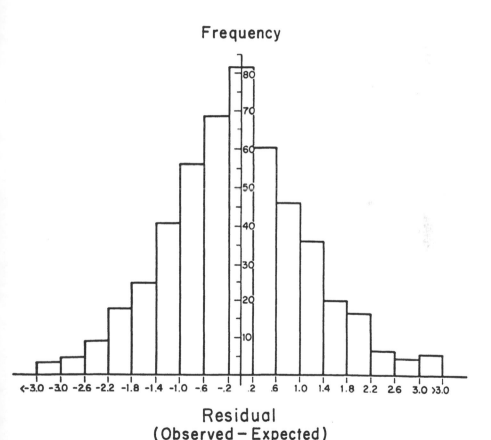

Figure 5.3. Distribution of residuals of two-step transitions. The horizontal axis is given in units of the number of transitions from square A to B to C. A positive residual indicates more observations of these transitions than expected on the basis of the transition probabilities.

126

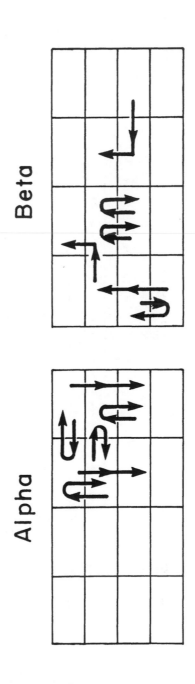

Figure 5.4. Zones of attraction for alpha and beta. The arrows identify the two-step transitions that occur far more frequently than expected (residual of +2.5 or more). The zones of attraction are associated more commonly with the eggs than expected for alpha and less commonly than expected for beta.

aberrant two-step transitions occurring within the egg-squares (162/530 = 0.306 is overall frequency of occurrence within egg squares versus 2/18 = 0.11 for the frequency of aberrant two-step transitions, t = 2.05, $\underline{P}$ < .05).

The movement patterns of the ants within the nest are not explained by a Markov movement rule. There are specific transitions that are far more likely than expected. These transitions are localized into zones of attraction for alpha and beta. The zones of attraction are associated with egg squares more strongly than expected for alpha and less strongly than expected for beta.

Movement of ants is a function of the position of the eggs interacting with the rank of the ant. Movement is also affected by relative rank of ants that encounter one another within the nest. When a moving ant encounters a stationary ant within the nest it may have one of several reactions. It may detour only slightly in its progress through the nest and move around the stationary ant. Alternatively it may turn completely around and return in the direction from which it came. Much of the time it will do something in between.

Whether the ant alters its movement only slightly or dramatically depends upon the relative hierarchy ranks of the two ants, as shown in Table 5.3 for movement that was not associated with dominance activity. If the higher ranking ant is stationary, the moving ant detours substantially and ends by returning in the direction from which it came. If the higher ranking ant is moving it continues on its course after merely detouring around the stationary ant.

Space Use

The pattern of space use during the quiescent phase of the colony activity cycle is shown for a number of ants in Figure 5.5. The pattern of space use is significantly different from the pattern of occupancy in the movement data ($\underline{P}$ < .05, X^2 for all cases). The pattern of space use is significantly different from the pattern of egg occurrence (X^2 for all cases, $\underline{P}$ < .05), although

Table 5.3
Movement detours as a function of relative
rank

	Substantial Detour	Slight Detour
Stationary Ant Higher in Rank	12	2
Stationary Ant Lower in Rank	0	9

Fisher's Exact Test _P_ < .01.

there is a significant positive rank-correlation
between egg occurrence and space use for some ants
(see Table 5.4).
     The pattern of space use is clearly a
function of hierarchy rank. Alpha, in either
colony, is found in close association with the
eggs during the quiescent phase of colony
activity. This is reflected in the average
distance of an ant from the eggs shown in Figure
5.6. Alpha is within 2 mm of the eggs, whereas
beta is much further, on average, from the eggs.
In one colony gamma is much closer to the eggs
than beta; in the other colony the distances of
beta and gamma are similar.
     On a more microscopic scale much the same
pattern is observed. Figure 5.7 shows the
distance of a given ant from eggs given in body
lengths. When an ant is zero body lengths from
the eggs it is standing upon them. Alpha is found
most frequently standing upon the eggs. Beta is
not usually found in such close proximity to the
eggs. However, gamma may be found with greater
frequency within one body length of the eggs.
Workers of rank of fourth or lower are found
almost exclusively at greater than one body length
from the eggs. Alpha can, in practice, be
tentatively identified by observing the colony
during the quiescent phase as the ant that is most

0	0	.09	.05
0	.02	.07	.33
0	0	.16	.23
0	.02	.02	.02

α - 798

.02	.07	.04	.02
.05	.09	.04	.16
.05	.09	.05	.09
.09	.05	.02	.09

β - 798

0	0	.14	.14
0	.04	.04	.23
0	.02	.05	.16
0	.04	.04	.09

γ - 798

0	0	.13	.04
.04	.15	.31	.19
0	.02	.04	.04
0	.02	.02	0

α - 795

.04	.04	.02	.02
.12	.10	.06	.12
.04	.02	0	.06
.12	.06	.08	.10

β - 795

.02	.08	.08	0
.02	.02	.04	.14
.08	.16	.02	.06
.08	.02	.04	.02

γ - 795

Figure 5.5. Pattern of space use for the three highest ranking ants of two colonies. The numbers give the proportion of time during the quiescent phase in which a given ant was in a given square.

Table 5.4
Rank correlation between space use and
egg occurrence

Colony	Ant	Correlation
798	α	0.696[a]
	β	-0.230
	γ	0.535[a]
795	α	0.673[a]
	β	-0.025
	γ	0.807[a]

[a] $p < .01$ (two-tailed).

closely associated with the eggs.

As can be seen from Figure 5.5, certain ants, for example α 798 and α 795, are very specific about the squares that they occupy during the quiescent phase. Other ants are found in a number of squares during this period. One can express this tendency as a measure of positional diversity. Those ants which occupy many squares have a high positional diversity; those ants which occupy only a few squares have a low positional diversity.

Positional diversity should be measured by the inverse of the variance of position. Since there is a slightly variable number of observations for each ant, the positional diversity is given as the reciprocal of the variance times the mean.

When a member of the hierarchy shows some sort of site affinity it is the eggs that the ant clumps around. Figure 5.8 shows the relationship between positional diversity and the mean distance to the eggs. When the ants are very clumped the distance to the eggs is small; when the ants are far from the eggs the positions are not clumped. For ants that are not members of the hierarchy, there is no relation between positional diversity and the distance to the eggs. The deviant point

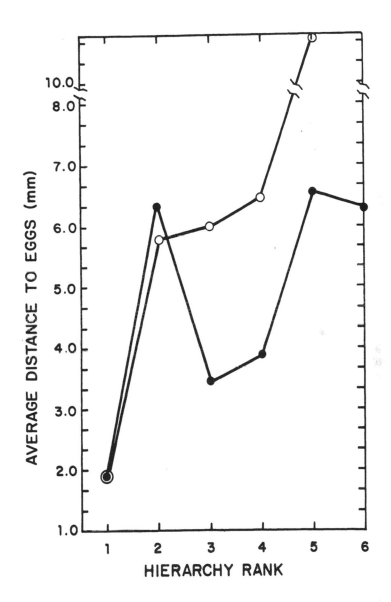

Figure 5.6   The average distance of an ant to the
      eggs during the quiescent phase of the colony
      activity cycle as a function of hierarchy
      rank.   The relation for two colonies is
      illustrated.   Solid dots = colony 795; hollow
      dots = colony 798.

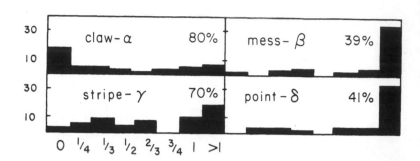

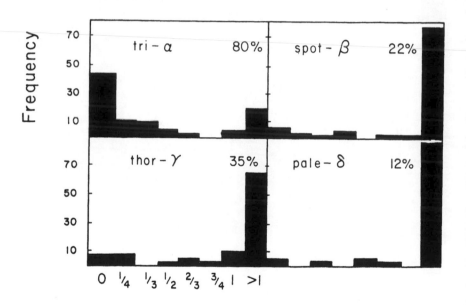

Distance to Eggs

Figure 5.7. The distance to the eggs of the three
highest ranking worker ants in each of two
colonies during the quiescent phase of the
colony activity cycle. The units are in ant
body lengths. A body length of zero
indicates that the ant is standing on the
eggs. The percentage in upper right corner
is the percentage of occurrence within one
body length of the eggs. Top panel = colony
795; bottom panel = colony 798.

among hierarchy members in Figure 5.8 is due to the lowest ranking member of one colony who became unranked during the period of study.

One can calculate a similar diversity measure for the movement data of Figure 5.2. The relation between movement diversity and positional diversity is shown in Figure 5.9. Ants that have clumped movement patterns also have clumped spatial patterns during the quiescent phase. These two sets of data are independent of one another and, in fact, some ants show clumping patterns around different squares for movement patterns and spatial patterns.

One can measure the degree of association between movement patterns and the position of the eggs as:

$$C = \Sigma_i p_i q_i,$$

where p_i is the fraction of the movement transitions that take place in square i, and q_i is the fraction of the times that eggs are found in square i. The relation between movement diversity and movement-egg correlation is shown in Figure 5.10. While the two quantities are not as tightly correlated as positional diversity and mean distance to the eggs, it is nevertheless true that when movement diversity is lowest (movement localized) the correlation between movement and egg position is highest.

DISCUSSION

I shall address two major topics in this discussion: the organization of colony activity and the dependence of worker behavior on hierarchy rank. These topics are not unrelated. As I shall demonstrate here and elsewhere (Cole 1986), colony organization is the outcome of independently behaving workers, each acting so as to maximize its potential for reproduction.

Wilson (1971) described the principles of mass action governing complexity in social insects. Individual workers act independently and respond to stimuli of the moment. However, selection acts at the colony level with the result

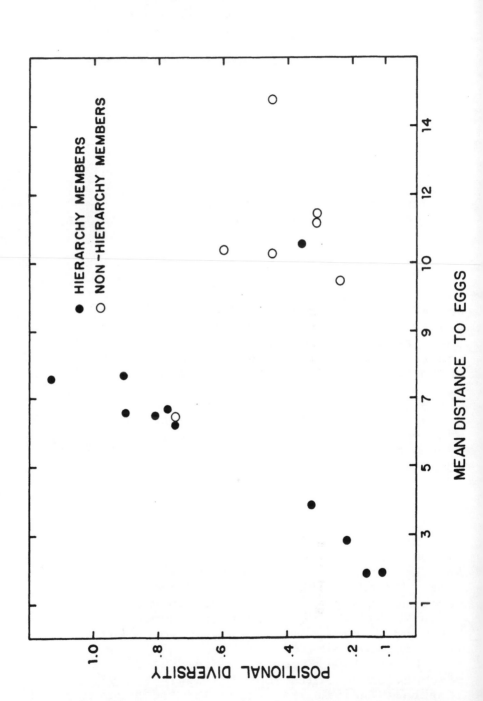

Figure 5.8. The relationship between positional diversity and the mean distance (mm) to the eggs. Hierarchy members show a positive relation. Non-hierarchy members show no relation.

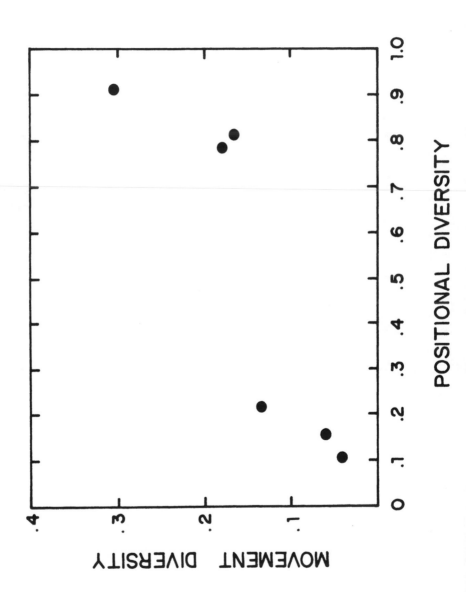

137

Figure 5.9. The positive relationship between movement diversity and positional diversity.

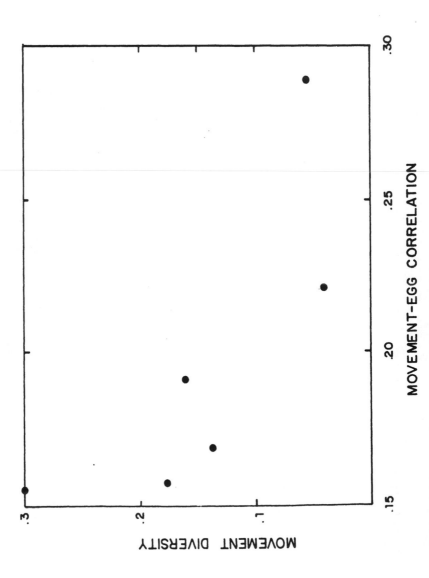

Figure 5.10. The relationship between movement diversity and the correlation between egg position and movement. When movement diversity is low (clumping of movement patterns) there is a higher correlation between movement and the position of the eggs.

that the behavioral program of the workers produces efficient colonies, tracks environmental change, and yields increasing colony efficiency.

In many species of ants this may be correct. For species in which worker ants do not have the capacity for reproduction the fitness of one worker is the same as that of any other. In such a case each worker should attempt to increase the efficiency of the colony. The results of the superorganism concept of the social insect colony are reclaimed, albeit for radically different reasons.

However, in many other species this is not completely accurate. When a worker ant has the capacity for reproduction it is no longer true that the fitness of each worker is identical. The worker that reproduces successfully has a different fitness from that of the worker that does not. This difference allows natural selection to operate between individual colony members and increased colony efficiency need not be the result. In L. allardycei this seems to be a more accurate picture. How does the behavior of individual workers influence the integration of colony function?

Understanding the production of the complexity of social insect behavior from the simple elements of behavior of individual workers is one of the challenges of social insect biology. Movement within the nest is one of the basic categories of worker behavior that is required for the care of brood, transport of food, and care of the queen. Individual worker movement brings workers into contact with brood, without which the brood are not fed, cannot moult, and do not eclose to adulthood. Individual worker movement brings workers together for oral trophallaxis, allogrooming, and dominance encounters. The queen depends on worker movement for nourishment, grooming, and to dispose of the eggs that she lays.

In L. allardycei, at least, the movement of workers is not accurately described by a Markovian movement model. Where an ant is going is a function not only of current position but of past movement and the rank of the ant. The fact that dominance interactions occur most frequently in

close proximity to the eggs is at least consistent
with a lack of independence of movement by pairs
of ants. It may also be explained by an increase
in aggressiveness near the eggs. Since the
explanations are not competing hypotheses, both
may, in fact, be correct. There are some
consequences for colony organization to be drawn
from these observations. First, workers do not
respond independently to stimuli of the moment.
That is, they deviate from the paradigmatic
principles of mass action. Second, individual
differences in movement are a function not only of
individual differences in activity levels but also
of individual differences in rank. Third, the
functions of the colony are accomplished in the
face of constraints on movement.

In the movement patterns of ants there is
evidently a memory of some sort. Movement to
certain parts of the nest is dependent upon past
movement. This makes more difficult the
description of overall colony functioning. Given
the paradigmatic principles of mass action, one
can deduce a sort of overall statistical mechanics
for an ant colony. In L. allardycei, while there
is clearly order, it is not assembled from the
concatenation of interachangeable parts. It is
not possible to extrapolate movement patterns
within the nest from the movement patterns of a
representative ant; there is no such individual.
It is uninteresting to model the movement patterns
of ants within the nest by recourse to some
phenomenologically fitted "average worker."

The orderliness of movement in L. allardycei
colonies is a function of the attraction of the
brood and the agonistic behavioral interactions
that result in hierarchy rank. Aggression between
workers, particularly the escalated aggression
between alpha and beta, is reflected in the
avoidance response of moving ants to stationary
ants of higher rank (Table 5.3). Various aspects
of the movement pattern, including the differences
in occurrence frequency (Figure 5.2) and
differences in the zones of attraction (Figure
5.4) between alpha and beta are probably the
result of movement corresponding to agonistic
interactions that occur most commonly near the
eggs (Table 5.1). Movement is dependent not only

on rank but also on the position of the egg pile. This is illustrated by the fact that the zone of attraction of alpha is highly associated with egg-squares (Figure 5.4) and the fact that high movement diversity is associated with low movement-egg correlation (Figure 5.10).

The impression left by the influence of individual interactions on movement patterns is not one of increasing efficiency. The aggressive interactions of workers appear to result in movement patterns altered by constraints. The notion that one worker actively interferes with the movement patterns of another worker is inconsistent with selection at the colony level to increase colony efficiency. Rather, it is consistent with the idea that each worker is behaving in a manner so as to come into proximity with the eggs. That such behavior is the result of selection on workers to maximize their individual fitnesses shall be argued below and elsewhere (Cole 1986).

The pattern of space use is also the result of the differing response of differently ranked ants to the position of the eggs. Alpha is found near the eggs, often standing upon them. Gamma is usually in closer proximity to the eggs than is beta, who is often at some distance from the pile. In addition the analysis of positional diversity shows that when positions are clumped it is the eggs that the ants clump around. In this context it is especially interesting to note that workers that are not members of the hierarchy may have some sort of site affinity, but it is not associated with eggs.

For space use, as for movement patterns, the data are better interpreted in light of constraints imposed by hierarchy rank than by the notion of efficiency governing colony function. Agonistic interactions result in a space use pattern that reflects an underlying inefficiency in the colony. If certain workers are excluded from certain areas of the nest by the aggression of other workers, the colony as a whole must become less efficient.

In L. allardycei workers are capable of reproduction. While the workers are uninseminated, they can produce unfertilized, male-producing eggs. The highest ranking members of the hierarchy have the most highly developed ovaries and are evidently responsible for the bulk of the reproduction by workers.

Much of the movement patterns and space use patterns can be interpreted as an attempt by alpha to monopolize reproduction. First, escalated conflict occurs more frequently between alpha and beta than between alpha and gamma. Second, most dominance encounters take place near the eggs. Third, more of alpha's movements occur in the vicinity of the eggs than do beta's. Fourth, the transitions that account for the deviations from a Markov movement pattern in alpha occur in even greater association with the eggs. For beta this zone of attraction is significantly less associated with the eggs. Fifth, there is a high rank correlation between the position of the eggs and areas of frequent movement. Sixth, alpha is found in close association with the eggs, frequently standing on top of them. Seventh, there is a close correlation between the positional diversity of an ant and its mean distance to the eggs.

Reproductive conflict of interest results in aggression between individuals. The reproductive asymmetry associated with rank exacerbates the conflict of interest. The routine occurrence of aggression between colony members, reflected in the existence of a dominance hierarchy, and the escalation of conflict between alpha and beta illustrate the conflict of interest between individual members of the colony. Not all workers have the same reproductive success, as illustrated by differences in ovarian development.

Personal reproduction by workers is not achieved without a cost. The constraints placed on movement and space use patterns as a function of hierarchy rank, which is the outcome of a reproductive conflict of interest, must reduce efficiency of colony functions. The extent to which reduced efficiency can be tolerated by reproducing ants is explored more completely elsewhere (Cole 1986).

SUMMARY

There are four reasons that one may conclude that there are individual differences in social insect behavior: inappropriate comparison, statistical variability, colony-level modification of individual behavioral ontogenies, and individual fitness differences. I examine the fourth cause of individual fitness differences in behavior for movement and space use patterns in _Leptothorax allardycei_. Dominance encounters occurred more frequently in proximity to eggs and encounters between alpha and beta were more frequently escalated. Ants move in patterns that put them in close proximity to the eggs. Movement is non-Markovian and differs between alpha and beta. Patterns of space use are a function of hierarchy rank. Alpha becomes quiescent in close proximity to the eggs, often standing upon them. There is a correlation between positional diversity and the distance to the eggs for hierarchy members. Patterns of movement and space use are consistent with the hypothesis that worker ants are attempting to maximize their personal reproduction even at the cost of reduced colony efficiency.

REFERENCES

Cole, B.J. 1981. Dominance hierarchies in _Leptothorax_ ants. _Science_ 221:83-84.

Cole, B.J. 1986. The social behavior of _Leptothorax_ ants: time budgets and the evolution of worker reproduction. _Behavioral Ecology and Sociobiology_ 18:165-173.

Jeanne, R.L. 1972. Social biology of the Neotropical wasp _Mischocyttarus drewseni_. _Bulletin of the Museum of Comparative Zoology, Harvard University_ 144:63-150.

Strassmann, J.E. 1981a. Evolutionary implications of early male and satellite nest production in _Polistes exclamans_ colony cycles. _Behavioral Ecology and Sociobiology_ 8:55-64.

Strassmann, J.E.   1981b.   Wasp reproduction and
    kin selection--reproductive competition and
    dominance hierarchies among Polistes
    annularis foundresses.   Florida Entomologist
    64:78-88.
West, M.J.   1967.   Foundress associations in
    polistine wasps: dominance hierarchies and
    the evolution of social behavior.   Science
    157:1584-1585.
West Eberhard, M.J.   1969.   The social biology of
    polistine wasps.   Miscellaneous Publications,
    Museum of Zoology, University of Michigan,
    140:1-101.
West Eberhard, M.J.   1977.   The establishment of
    reproductive dominance in social wasp
    colonies.   Proceedings of the 8th
    International Congress, International Union
    for the Study of Social Insects, Wageningen,
    The Netherlands.   pp. 223-227.
Wheeler, W.M.   1928.   Foibles of insects and
    men.   New York: Knopf.
Wilson, E.O.   1971.   The insect societies.
    Cambridge, Massachusetts: Harvard University
    Press.

6

The Gyne Who Would Be Queen:
Dominance in the Ant *Iridomyrmex purpureus*

Norman F. Carlin

Among the growing number of complex adaptive behavior patterns exhibited by social insects in which individual variability has been documented, dominance interactions between reproducing or potentially reproducing nestmates have perhaps the most direct effect on fitness. Dominance hierarchies and their reproductive consequences have been investigated in a variety of wasps and bees characterized by varying levels of sociality (Pardi 1948; Röseler 1965; Jeanne 1972, 1980; West 1967; Brothers and Michener 1974; West-Eberhard 1977, 1978; Gadagkar 1980). Pheromonal dominance in the form of functional monogyny in polygynous species (and, of course, the inhibition of worker oviposition by queens) is well known throughout the social insects (Wilson 1971; Fletcher and Ross 1985). Behavioral dominance in highly eusocial ants has been relatively little studied by comparison. Hierarchies have recently been discovered among workers of some species, determining whether and which will lay male eggs (Cole 1981; Franks and Scovell 1983). Coexisting queens in pleometrotic or polygynous ant species are only beginning to be scrutinized for dominance interactions (Hölldobler and Taylor 1983; Fowler

The data in Tables 6.2, 6.3 and 6.5, and some other results and discussion presented here, have previously been published by Hölldobler and Carlin (1985) in Behavioral Ecology and Sociobiology.

and Roberts 1983; Evesham 1984), and these studies
have not reported the effect of status on
reproduction.

Queens of the Australian meat ant,
Iridomyrmex purpureus (Smith), readily form
pleometrotic associations, and cofoundresses
housed together in the laboratory engage in
apparent dominance behavior in the form of
frequent, stereotyped bouts of rapid mutual
antennation. As queens of this species oviposit
at a rather high rate, laying up to 7 eggs per
hour, the effect of dominance on their
reproductive output is easily assessed. In the
present paper, detailed analysis of mutual
antennation behavior between a pair of associating
I. purpureus queens demonstrates that these
ritualized agonistic displays do in fact
constitute dominance contests, resulting in a
reproductive hierarchy in which egg-laying by the
subordinate is inhibited.

I. purpureus is an ecologically important
dolichoderine ant, widespread in Australia,
building large mound nests containing up to 60,000
workers (Ettershank 1971; Greaves and Hughes
1974). Most colonies are founded by solitary,
claustral queens following nuptial flights, i.e.
by haplometrosis (Greaves and Hughes 1974).
However, in a population near Mount Ainslie,
Australian Capital Territory, about 10 percent of
founding nests were found to be pleometrotic, with
2-4 queens each (Hölldobler and Carlin 1985).
Multiple foundresses may continue to coexist after
workers eclose, and mature nests can contain more
than one ovipositing queen (Greaves and Hughes
1974; Hölldobler and Carlin 1985). Unlike the
many queens of typically polygynous ants, which
tend to cluster amicably within the same nest
chamber, I purpureus queens in mature colonies are
few in number, located in widely separated nest
chambers, and are mutually aggressive when placed
together, though all workers in the colony accept
them. These traits are characteristic of
oligogyny, a state intermediate between monogyny
and polygyny (Hölldobler 1962).

Hölldobler and Carlin (1985) found that queen
dominance is the mechanism by which oligogyny
develops and is maintained in this species. As

the colony increases in size, the cofoundresses become progressively less tolerant of one another's presence, but do not fight. Instead, they move into separate chambers within the nest, and are equivalent in status thereafter, laying equal numbers of eggs. Since the queens have long since separated by the time the colony has reached a size of several thousand workers and begun to rear reproductives, the selective advantage of costly dominance interactions during the early stages of colony development is not obvious. Possibly the dominant queen attempts to reduce later competition over alate production by permanently subordinating her cofoundress while they are still together; lasting inhibition following the experience of low status is known in some other social insects (Wheeler 1986).

METHODS

Founding queens of _Iridomyrmex purpureus_ were collected from the Mount Ainslie area near Canberra, Australia, in October 1980 by B. Hölldobler. Foundresses were housed in the laboratory, kept at 20-25°C, in test tube nests (15 cm x 2.2 cm) containing water trapped at the end behind a cotton plug. Pairs and groups of four queens, collected from different localities and almost certainly unrelated, were housed together. Those kept in groups of four exhibited occasionally lethal fighting after workers eclosed (Hölldobler and Carlin 1985), while those in pairs formed stable associations. Once workers had eclosed, each nest tube was transferred to a plastic box whose walls were coated with Fluon (Northern Products, Inc., Woonsocket, R.I.), in which synthetic ant diet (Bhatkar and Whitcomb 1970), honey-water, and chopped cockroaches (_Nauphoeta cinerea_ [Olivier]) were provided twice weekly.

Five pairs of cofoundresses were observed for approximately 1 year. The queens in one of these colonies were marked on the thorax using Testor's PLA® paint; they received dots of blue and yellow, respectively, and will be referred to as queen B and queen Y. As this focal colony grew to contain

hundreds, later thousands of workers over a two
and one-half year period, three more plastic boxes
filled with water tubes were provided in addition
to the original nest box. The boxes were set
inside a large foraging arena (152 x 76 cm), to
which each was connected by a cardboard bridge.
While the marked queens were together, all
queen-queen and queen-worker interactions and
solitary queen behavior were recorded during two
1.5-hour sessions daily, for a total of 74.5 hours
of observation. Data was collected during the day
and at night, as this appeared to have no effect
on queen behavior, though workers foraged
diurnally. (The colony was illuminated 24 hours
per day during the weeks when observations were
made.) For each antennation bout, the following
were recorded: the identity of the queen(s) which
initiated and terminated antennation, whether
antennation was reciprocated, the duration of the
bout, whether either queen retreated at the end of
the bout, the locations of the queens, and other
behavior occurring during or immediately following
(within 5 seconds of termination) the bout--e.g.,
fighting or oviposition. Bout durations were
timed to within 0.1 second with a digital
stopwatch. An encounter was only considered to be
a fully mutual bout if reciprocal antennation
lasted more than 0.5 seconds; if the antennated
queen did not respond, or gave just a few brief
twitches of her own antennae, the bout was scored
as "not reciprocated". Queen locations during and
outside of bouts were recorded with respect to the
central pile of brood, containing mostly eggs.
Rates of oviposition by the two queens after they
had separated permanently were noted during 3
hours' observation of each.

RESULTS

Until workers began to emerge, 2.5 months
after founding, queens remained in their nest
tubes without feeding, laid eggs and cared for
brood, and did not interact in any obvious way.
After the first workers eclosed, cofoundresses
began to perform stereotyped bouts of rapid mutual
antennation. Among pairs of queens, antennation

escalated only rarely into threatening with wide-open mandibles and biting. Cofoundresses engaged in antennation bouts, each lasting about 2 seconds, at an average rate of 30 times per hour for nearly a year; this must represent a considerable investment in time and energy. The queens rapidly fenced with their antennae, flagellum-to-flagellum, flagellum-to-scape or flagellum-to-head (aiming particularly at the mouth region), as if each was trying frantically to solicit regurgitation from the other. Queens never antennated workers in this manner. Though they used antennation to solicit regurgitation and trophic eggs from workers, and the latter often briefly reciprocated antennation before offering food, the rate of antennal movement was always much slower and more regular than in queen-queen bouts. Frequently a bout began when two queens solicited the same worker for food, the worker then moving out from between them. However, actual trophallaxis between queens was observed only once. Thus, though bouts seem to represent a ritualized version of the antennal movements used in solicitation (as do dominance displays in some wasps [Jeanne 1972]), they do not now function in food exchange. Workers completely ignored all queen-queen antennation bouts.

Both members of a pair initiated antennation bouts. All bouts during the period before separation ended with one queen or the other unambiguously backing or turning away to terminate the encounter. The rate of antennation tended to increase late in each bout, just before the terminating queen broke off. The pair often immediately re-engaged in another bout, or walked away from one another, or groomed themselves. Queens stood still at antenna's length while antennating, their mandibles closed, partly open, or opening and closing repeatedly. Usually they were about 1-2 mm apart during bouts, but occasionally their mandible tips came into contact. Sometimes they faced one another while antennating, or sometimes stood side by side, facing in the same direction, with their heads turned sideways toward one another. Bouts did not necessarily occur each time queens met; they often stood close together or in physical contact without antennating. Antennation initiated by one queen frequently was not reciprocated, the recipient tending not to react if antennated on thorax or gaster, or if otherwise engaged in feeding, self-grooming or being groomed. While being groomed by workers, a queen often appeared to "go into a trance," rolling on her side or back and lying motionless in a contorted posture for up to 5 minutes, during which she was unresponsive to antennation by the other queen. Queens never groomed one another.

After bouts had begun, each pair remained together in its original founding nest tube, though additional tubes were provided to accommodate increasing numbers of workers. After about a year, when the colonies contained 200-400 workers each, antennation bouts became more frequent, and the queens began to briefly leave the founding tubes. Fourteen months after founding, both queens in the focal colony, which had been provided with additional nest boxes in a large arena, were observed to cross the cardboard bridges and enter new nest boxes. These separations lasted for increasing periods from several hours to 33 days, but the queens intermittently rejoined one another in their original box, and resumed antennation. Nine months later, the queens separated permanently, by which time the colony contained about 4500 workers and was producing males. These stages in colony ontogeny were referred to as phase I (before separation), phase II (incipient separation, with intervals of reunions), and phase III (after final separation) by Hölldobler and Carlin (1985).

Focal Colony Observations

Dominance Before Separation. The two individually-marked queens in the focal colony, queen B and queen Y, appeared to be equal in size, though they were not weighed. Sixty hours of observation were taken over a period of 4 weeks in mid- to late phase I, when the queens were venturing infrequently out of the founding tube for not more than 3 minutes at a time. A total of 1908 queen-queen encounters were recorded; in 1335 bouts, the pair engaged in mutual antennal fencing lasting more than 0.5 seconds. The duration of mutual bouts was 1.9 ±2.3 seconds (mean ±SD), with a range of 0.5 to 19.9 seconds.

Every reciprocated bout before separation ended when one queen or the other unambiguously backed or turned away. Queen Y terminated encounters more often than did queen B, in 84 percent of 1335 bouts, while queen B terminated only 16 percent; this difference was highly significant (Table 6.1). Queen B laid 430 eggs during observations (excluding ovipositions during bouts, which will be discussed below), while queen Y laid only 215, also a highly significant difference (Table 6.1). The low frequency of terminating bouts by queen B coupled with her higher oviposition rate, and supported by other correlations reported below, suggest behavioral and reproductive dominance. For this reason, mutual bouts which one queen terminated will be referred to below as having been "won" by the other; accordingly, queen B was the winner and queen Y the loser in most bouts.

Queen Y initiated significantly more mutual bouts than did queen B (Table 6.1), though the durations of bouts initiated by the two queens were not significantly different (Table 6.2). Initiating antennation does not seem to be a meaningful indicator of status in this species, as the probability of winning a bout was not significantly affected by having initiated it (Table 6.3).

Initiation, unlike termination, was uncorrelated with oviposition rates or the other queen acts of Table 6.3. The pair remained close together at the end of most encounters, but the loser walked away for a distance of more than two queen-lengths after 79 bouts. (The winner never retreated.) Queen B, consistent with her dominant rank, retreated after terminating far less often than did queen Y (Table 6.1). The occurrence of non-reciprocated antennation was also consistent with queen B's status; she ignored antennation by queen Y significantly more than the latter ignored her initiation (Table 6.1). The two queens did not differ significantly in self-grooming behavior, which provides a control for possible

154

Table 6.1
Number of antennation bouts and other behavioral
acts performed by two individually-marked queens
in a colony of Iridomyrmex purpureus

Act	Queen B	Queen Y	$\underline{P}$[a]
	Before separation		
Initiate bout	618	717	<.01
Terminate bout	215	1120	<.001
Walk away	9	70	<.001
Ignore antennation	406	167	<.001
Self-groom	600	662	NS
Lay egg			
outside of bouts	430	215	<.001
during a bout	6	21	<.005
	During reunions		
Initiate bout	145	139	NS
Terminate bout	145	99	<.005
Walk away	13	6	NS
Ignore antennation	50	40	NS
Self-groom	167	194	NS
Lay egg			
outside of bouts	99	75	NS
during a bout	4	3	NS
	After separation		
Lay egg	21	26	NS

[a]Differences are considered significant at
$\underline{P} < .05$ (x^2 test)

variation in general activity levels that might
affect dominance (Table 6.1).

Queens were not observed eating one another's
eggs, though they continually consumed workers'
trophic eggs. (Oophagy is an important part of
the dominance repertoire of a number of
Hymenoptera [Fletcher and Ross 1985].) Both
ignored even eggs laid right in front of their
mandibles, while antennating the other from
behind. On three occasions a queen ate an egg
from the egg pile; as workers were often seen
placing their own trophic eggs on the pile, these
may not have been produced by queens.

Dominance During Reunions. The ranking
observed in phase I, before the first separation,
did not persist; when the queens reunited during
the incipient separation period, phase II, their
interactions were significantly altered. Both
queens left the original founding tube at the
beginning of this phase, but surprisingly, it was
dominant queen B which finally began to settle in
a new nest box. In her absence, queen Y
established herself in the original box. A total
of 374 antennation bouts were recorded in 14.5
hours of observations during 4 reunion intervals,
each lasting 1-4 days, in the second month of
phase II. The first two reunions followed
intervals in which queen B had been residing in
another tube in the original nest box, while the
latter two followed intervals when queen B had
been in another nest box. Three reunions took
place in the founding nest tube, but during the
last the pair encountered one another in a
neighboring tube in the original nest box.

Of 244 mutual antennation bouts recorded
during phase II in which one queen unambiguously
terminated, queen B won only 41 percent, less than
half her success rate in phase I. Queen Y laid 75
eggs during reunions, compared to 99 laid by queen
B, an insignificant difference (Table 6.1;
excluding ovipositions during bouts). During 6
hours of observation in the third month of phase
III, after the queens permanently separated, they
laid equal numbers of eggs (21 by queen B, 26 by
queen Y; Table 6.1). The reversal in the
frequencies of terminating bouts and the increase
in ovipositions by queen Y, once she was freed

Table 6.2
Effect of other queen behavior on duration of mutual antennation bouts before separation and during reunions

Queen act during bout	Bouts in Which Act is Performed		Bouts in Which Act is not Performed		
	Duration(s)[a]	N	Duration(s)[a]	N	P[b]
Before separation					
B wins bout	1.7 ±2.2	1120	2.8 ±2.7	215[d]	<.001
B initiates	1.9 ±2.5	618	1.9 ±2.2	717[e]	NS
B won previous[c]	1.9 ±2.3	1262	2.1 ±1.7	70	NS
Y won previous[c]	2.1 ±2.5	517	1.8 ±2.1	815	<.005
B lays egg	2.0 ±2.2	6	1.9 ±2.3	1329	NS
Y lays egg	3.4 ±3.4	21	1.9 ±2.3	1314	<.01
B on eggs	2.0 ±2.4	1129	1.6 ±1.7	206	<.05
Y on eggs	1.9 ±2.5	775	1.9 ±2.1	560	NS

			During reunions		
B wins bout	4.9 ±11.3	99	2.7 ±5.3	145[f]	<.05
B initiates	2.6 ± 5.8	145	3.7 ±9.3	139[e]	NS
B won previous[c]	4.4 ± 9.5	143	1.8 ±5.3	139	<.01
Y won previous[c]	3.3 ± 8.4	221	2.5 ±4.7	61	NS
B lays egg	2.4 ± 3.8	4	3.2 ±7.8	280	NS
Y lays egg	2.9 ± 4.2	3	3.1 ±7.8	281	NS
B on eggs	3.6 ± 7.6	153	2.6 ±7.9	131	NS
Y on eggs	4.1 ± 9.6	177	1.5 ±1.5	107	<.01

[a] Durations are means (seconds) ± standard deviation. N = total numbers of bouts during which each act is performed or not performed.

[b] Differences are considered significant at P <.05 (one-way analysis of variance).

[c] Won at least 1 of the 5 previous bouts.

[d] Bouts before separation that were not won by B were won by Y.

[e] Bouts that were not initiated by B were initiated by Y.

[f] Number of bouts observed totals 244 for "wins bout," 284 for all other acts, because 40 bouts with no unambiguous winner were observed during reunions.

Table 6.3
The effect of other queen behavior on the probability of winning mutual antennation bouts before separation and during reunions

Queen act during bout	Bouts Won in Which Act is Performed		Bouts Won in Which Act is not Performed		p^b
	Proportion	N	Proportion	N	
Before separation					
B initiates	0.824	618	0.852	717[c]	NS
B won previous[a]	0.851	1262	0.643	70	<.001
Y won previous[a]	0.238	517	0.110	815	<.001
B lays egg	0.667	6	0.840	1329	NS
Y lays egg	0.762	21	0.151	1314	<.001
B on eggs	0.857	1129	0.737	206	<.001
Y on eggs	0.169	775	0.150	560	NS

			During reunions		
B initiates	0.357	129	0.461	115[c]	NS
B won previous[a]	0.647	136	0.094	106	<.001
Y won previous[a]	0.736	182	0.167	60	<.001
B lays egg	0.750	4	0.400	240	NS
Y lays egg	1.00	3	0.589	241	NS
B on eggs	0.475	139	0.314	105	<.05
Y on eggs	0.745	150	0.500	94	<.001

A bout is considered won by a given queen when the other terminates. Proportion of bouts lost in which a given act is performed = 1 - proportion of bouts won. N = total numbers of bouts during which each act is performed or not performed.

[a] Won at least 1 of the 5 previous bouts.

[b] Differences are considered significant at $P \leq .05$ (χ^2 test).

[c] Bouts that were not initiated by B were initiated by Y.

from interactions with queen B, together demonstrate that bouts do in fact represent dominance interactions. From founding to the first separation, queen B won most bouts and inhibited egg laying by queen Y, while the latter was disinhibited, her oviposition rate increasing, after separation.

Early in each reunion, the queens tended to avoid one another, engaging in a few brief 1-2 second antennation bouts, but later bouts increased in duration and intensity. Often the pair engaged in sequences of 2-4 bouts in rapid succession, each shorter than the last, before breaking off; this pattern had not occurred before separation. Each queen's antenna tips contacted the other's clypeus more frequently during reunion bouts. It appeared that in later reunion encounters queen B was beginning to regain her dominance, only to lose it again following another long absence. Once, late in a reunion interval, when queen Y briefly left the founding tube, queen B stood in the tube mouth and, in a long (14 second) bout of very rapid antennating, prevented queen Y from reentering. Several minutes later queen B moved back to the brood pile and queen Y reentered.

There was no significant difference between the numbers of bouts initiated by queen B and by queen Y during reunions (Table 6.1). As in phase I, which queen initiated had no effect on bout duration (Table 6.2) or on the likelihood of winning (Table 6.3). Queens walked more than two queen-lengths away following bouts on too few occasions for significance, though queen B retreated twice as often as did queen Y (Table 6.1). Consistent with queen B's loss of status, there was now no difference in the numbers of antennations ignored by either queen (Table 6.1), and their rates of self-grooming still did not differ (Table 6.1).

Effect of Winning on Dominance Motivation. While the pair coexisted prior to separation, queen B appeared to have become accustomed to her dominant status. Bouts won by queen Y were significantly longer than those won by queen B (Table 6.2). The rate of antennation by queen B during longer bouts also was somewhat higher than usual. Thus queen B seemed to be a "sore loser," disputing her losses with prolonged antennation, while queen Y gave up more readily. Only when the latter continued antennating longer was she able to win.

Separation and the resulting destabilization of status apparently made both queens less ready to terminate a bout. All bouts during reunions were significantly longer and more variable in duration than those before separation, lasting 3.6 ± 8.3 seconds (x ± std. dev.), range 0.5-75.2 seconds compared with 1.9 ± 2.3 seconds in phase I (P < .005 by one-way analysis of variance). In addition, though most mutual bouts during phase II ended with one queen unambiguously terminating (as had all bouts in phase I), 40 ended when the rate of antennation slowed and gradually stopped, neither queen backing off. Queen Y also became the "sore loser" during reunions, when she became the resident and queen B the intruder. Reunion bouts won by queen B were significantly longer than those won by queen Y (Table 6.2), as the latter was less willing to give in, and the former terminated more easily. Thus in each phase, the currently subordinate or non-resident queen had to contest a bout for longer in order to win.

On the other hand, in both phases, both queens exhibited additional motivation to win a bout if they had won recent encounters. Having won one or more of the previous five bouts, each was significantly more likely to win the current one than if she had lost all the preceding five (Table 6.3). Before separation, whether queen B had won at least one of the previous five bouts did not affect the duration of the current one, but the current bout was significantly longer if queen Y had won at least one than if she had lost all five (Table 6.2)--the subordinate queen Y did not give up as readily when encouraged by a win. During reunions, this difference was reversed: queen Y's winning or losing streaks did not affect current bout duration, while the current bout was significantly longer if queen B won at least one of the five preceding. A subordinate queen's successes were apparently intimidating to her rival, or encouraging to herself, or both. Thus each queen's immediate motivational state was altered based on recent experience, but this change was superimposed on the generally higher, long-term dominance motivation of the reigning "sore loser."

Effect of Oviposition During Bouts. The act of ovipositing during a bout, presumably also influencing the queen's motivational state, greatly increased the probability of winning for queen Y before separation. Queens either stood stiffly, with gaster extended, or bent the gaster forward toward the head while laying an egg. Both were observed jerking or rocking forward and back during many ovipositions, which may have contributed to the effect on winning (though they performed jerking motions just as often when ovipositing outside bouts). Queen Y won 16 of the 21 bouts during which she laid an egg (76 percent), but won only 15 percent of those bouts in which she did not oviposit (Table 6.3). Bouts during which queen Y oviposited were also significantly longer (Table 6.2). Queen B laid eggs during only six bouts, of which she won four; though she too might have shown some effect of egg laying, the difference is not significant due to the small number of observations. Oviposition by queen B had no effect on bout duration in phase I, and there were too few ovipositions observed during phase II for any significant correlations (though queen Y did win all three reunion bouts in which she laid an egg; Table 6.3).

Both of the two occasions, early in phase I observations, when an antennation bout escalated into physical attack were associated with ovipositions. Once queen B lunged at queen Y's head three times, snapping her mandibles and antennating all the while, then immediately laid an egg. On another occasion, queen Y laid an egg while antennating queen B's gaster, jerking violently, then climbed on her back and nipped at her neck. In each case, the victim moved rapidly away and was not chased. No attacks occurred during reunions.

Since queen Y normally oviposited at a much lower rate than queen B, she laid significantly more eggs than expected during bouts (Table 6.1). It is conceivable that queen Y "deliberately" oviposited during bouts to obtain the temporary dominance advantage evident above. Nine percent of all her ovipositions in phase I occurred during an encounter (compared to 1 percent for queen B); assuming that the production of an egg requires some time, the stimulus of engaging in antennation seems unlikely to have caused so many simultaneous ovipositions, unless eggs already formed were retained for "use" during dominance interactions.

Effect of Location on the Egg Pile. As do many ant species, I. purpureus workers arrange brood at different developmental stages in more or less discrete areas within the nest. In laboratory colonies, small larvae were kept at the back of nest tubes (next to the cotton plug), large larvae and some pupae near the front, with a high heap of eggs and pupae near the middle; this is referred to as the "egg pile." To be queen of this small hill was another motivating factor and possibly an objective in dominance interactions. Before separation, dominant queen B spent 81.6 percent of her time atop the egg pile, while queen Y stood on the eggs only 40.7 percent of the time. Queen B defended her position with slightly but significantly longer antennation (Table 6.2), and was significantly more likely to win bouts while atop the eggs than while off (Table 6.3). Queen Y's location had no significant effect on bout duration or her probability of winning during phase I. This pattern was predictably altered during reunions, when resident queen Y became dominant. Queen Y spent 55.9 percent of her time on the eggs in phase II, compared with 49.6 percent by queen B. The location of queen B had no effect on bout duration, while queen Y's position on the egg pile increased duration dramatically (Table 6.2), and both queens had a higher probability of winning bouts while on the egg pile than while off (Table 6.3).

Location on the egg pile also affected the queens' oviposition rates. Dominant queen B, before separation, laid greater numbers of eggs than her rival while both were on the egg pile, while both were off, and when queen B was on but queen Y was off. However, queen Y laid more eggs than expected if she was on the pile while queen B was off (Table 6.4). In these respective

Table 6.4
Ovipositions by two I. purpureus queens while
located on or off the egg pile

Location	Queen B	Queen Y	P[a]
	Before separation		
Both on eggs	147	97	<.005
B on, Y off	221	91	<.001
Y on, B off	38	36	NS
Both off eggs	30	12	<.01
	During reunions		
Both on eggs	14	11	NS
B on, Y off	25	25	NS
Y on, B off	41	30	NS
Both off eggs	5	2	NS

[a]Differences are considered significant at $\underline{P}$ <
.05 (χ^2 test).

locations, any motivational stimulus the eggs
provide for queen Y, relative to queen B, is of
course maximized. The lack of significant
difference between the numbers of eggs laid is not
an artifact of small sample size, since even fewer
ovipositions occurred while both were off the
eggs, of which queen B nonetheless produced a
significant majority. During reunions, there was
no significant difference in ovipositions at any
queen locations. (The total number of eggs laid
during reunions in Table 6.4 is less than in that
given in Table 6.1, because in one reunion period
the queens met in a nest tube without a discrete
brood pile, and no location data could be taken.)
Queen Interactions with Workers and Brood.
In some pleometrotic ant species, workers exhibit
preferences for dominant queens, and aggression

toward subordinates, that eventually lead to the elimination of supernumerary cofoundresses (e.g., Myrmecocystus mimicus Wheeler [Bartz and Holldobler 1982]). However, in the I. purpureus focal colony, workers did not participate in their queens' dominance conflict, ignoring both queen-queen antennation bouts and queens moving from one nest tube to another. Neither queen received preferential antennation or grooming from workers in phase I, and more tellingly for queen reproductive success, neither received more regurgitated food and trophic eggs (Table 6.5). Queen Y did antennate workers significantly more frequently than did queen B; possibly the subordinate had to antennate more in order to be fed equally (Table 6.5). Queens very rarely groomed workers (each was observed doing so only once).

The change in dominance status during reunions was accompanied by alterations in queen-worker interactions. The queens still received equal amounts of food, but now intruding queen B antennated workers significantly more frequently than did resident queen Y. Workers developed a highly significant preference, antennating and grooming queen Y more than twice as much as they did queen B (Table 6.5). Though neither queen was fed preferentially at any time, on several occasions during phase II, queen B was seen soliciting from a worker, which responded by lightly nipping the tips of her mandibles instead of offering food. This mild antagonism was the only form of worker aggression toward either queen ever observed. Thus workers did discriminate in favor of the resident queen during reunions, but not in favor of the dominant one before they separated.

In phase I, both queens antennated the brood equally, but queen B performed significantly more handling and grooming of brood (Table 6.5). This might simply reflect her more frequent location atop the egg pile. Though larvae and pupae were present throughout the nest, eggs, which received most of the queens' brood-directed behavior, were available only on the central pile. "Handling and grooming brood" includes simply picking up and putting down clumps of eggs, a behavior queens

Table 6.5
Interactions between two I. purpureus queens and
their brood and workers

Act	Queen B	Queen Y	$\underline{P}$[a]
		Before separation	
Antennate worker	5131	5536	<.001
Antennated or groomed by worker	1410	1396	NS
Fed by worker	304	296	NS
Antennate brood	1499	1441	NS
Handle or groom brood	45	27	<.05
		During reunion	
Antennate worker	1261	1082	<.001
Antennated or groomed by worker	101	208	<.001
Fed by worker	47	62	NS
Antennate brood	214	302	<.001
Handle or groom brood	0	26	<.001

[a]Differences are considered significant at $\underline{P} <$ .05 (X^2 test).

performed more frequently than any other form of
brood "care." In addition to the behavioral acts
listed in Table 6.5, queen B was once observed in
trophallaxis with a larva, and queen Y once aided
several workers peeling the pupal skin off a
newly-eclosed callow.

During phase II, queen Y antennated, handled,
and groomed brood significantly more than queen
B. In fact, the latter did not once handle or
groom brood on those occasions when she returned
to the original tube (Table 6.5). The
contribution to total brood care by queens is
certainly negligible compared to that by

workers. However, it is conceivable that queen B ceased brood tending during reunions, after having moved out, because the brood she encountered in the founding tube was less likely to be hers. It is suggestive that before separation, when queen Y laid 35 percent of the eggs, she contributed 38 percent of brood care by a queen; during reunions, when she laid all the eggs in the pile at hand, she performed all the brood care. Adults of some ants have been shown to recognize related immatures (Lenoir 1984), and an I. purpureus queen may be able to discriminate her own brood from that of her cofoundress.

DISCUSSION

Additional field results support the laboratory evidence that dominance antagonism and the separation of pleometrotic pairs result in oligogynous colonies. Two of eleven mature nests excavated were found to contain two queens, separated in widely-spaced galleries within their respective colonies; in no nest were there more than two. Housed together in the laboratory, each pair of queens began performing mutual antennation bouts, and after some days moved apart. All four queens laid eggs, and dissection revealed that all had fully active ovaries (Hölldobler and Carlin 1985).

In other ant species exhibiting pleometrosis, the number of queens is usually reduced to one per colony before sexuals are produced (Hölldobler and Wilson 1977), often by fighting between the cofoundresses (Schmitz 1911; Waloff 1957; Rissing and Pollock 1986). Fighting also occurs in Iridomyrmex purpureus associations containing four queens (Hölldobler and Carlin 1985), but pairs engage in non-lethal dominance interactions instead. Rissing and Pollock (1986) propose that claustral cofoundresses are cooperative mutualists while the new colony is a "closed energy system;" when the first workers eclose, they constitute a monopolizable resource in an "open system," and conflict arises. This argument applies well to I. purpureus. Of 50 newly-mated queens housed in associations in the laboratory, none exhibited any torn of antagonism until workers emerged, after which overt or ritualized aggressive behavior began (Hölldobler and Carlin 1985).

Though in many species only one member of a foundress association survives to produce reproductives in the mature colony, pleometrosis is selectively advantageous for that survivor. Groups have a lower mortality rate than solitary queens, and produce more workers, in less time, in the initial brood which they rear without worker assistance (Bartz and Hölldobler 1982; Mintzer 1979; Rissing and Pollock 1986; Stumper 1962; Taki 1976; Tschinkel and Howard 1983; Waloff 1957). Pleometrotic groups of I. purpureus also produce significantly increased numbers of workers in their first brood (Hölldobler and Carlin 1985). But following the separation of cofounding queens, both contribute, presumably equally (assuming that the comparable oviposition rates observed early in phase III are maintained, as seems likely), to sexual production. Thus, I. purpureus queens enjoy the benefit of pleometrosis without suffering its usual cost.

The eventual success of both cofoundresses increases the advantage of joining associations. When pleometrosis leads to secondary monogyny, each of n associating queens has 1/n chance to inherit all (Hölldobler and Wilson 1977), but oligogynous queens, even those initially subordinate, have a better than 1/n chance of reproducing. Thus the usual requirement that 1/n must exceed the probability of successful haplometrosis, if unrelated foundresses are to form associations, is relaxed, with two possible implications. The number of eventually reproducing queens can increase--which it does not; lethal aggression occurs among four queens. Or, the probability of successful haplometrosis can be higher than in most pleometrotic species-- which it is, as 90 percent of I. purpureus colonies are founded solitarily. Since pleometrosis leading to oligogyny, with its advantageous head start for colony growth, has not in fact become the sole mode of founding, the likelihood of success by single foundresses must be reasonably good. Pleometrosis leading to secondary monogyny could not evolve under such

conditions, but oligogyny can. However, haplo-
and pleometrosis are not the only options open to
newly-mated I. purpureus queens. They may be
adopted by established colonies (probably their
natal nests), or by "satellite nests" that have
previously budded off from mature colonies
(Greaves and Hughes 1974; Hölldobler and Carlin
1985). The colony reproductive strategies of
these ants are quite complex, and merit further
study.

While cofoundresses remain together, their
ritualized aggression does represent a
disadvantage to the loser (that is, the
terminator) of antennation bouts by lowering her
oviposition rate. However, their ultimate output
of alates, long after separation has ended the
inhibition of egg production, is probably
similar. During the early stages in colony
ontogeny when dominance conflict actually takes
place, only workers are being produced. The
dominant queen invests about 1.7 percent of her
time for nearly a year (assuming 1.9 seconds/bout,
0.53 bouts/minute, from the eclosion of workers to
the first separation) and a considerable amount of
energy in subordinating her rival, to no obvious
adaptive purpose. If sperm were limited, and if
it were somehow behaviorally possible, each queen
would be expected to compete, not for the
privilege of producing sterile workers, but to
force the other to lay the worker eggs.

A possible explanation is that the dominant
queen, by inhibiting her cofoundress early on,
attempts to permanently subordinate the latter and
so to dominate the colony's eventual production of
alates even after the two separate. Dominance is
both a cause and a self-reinforcing effect of
reproductive variance in the social insects. A
hymenopteran female tends to become dominant
because she has well-developed ovaries and a high
titer of juvenile hormone (JH), whereupon she
monopolizes food resources and becomes even more
fertile. Her nestmates with less developed
ovaries and lower JH titers tend to become
subordinate; being dominated and inhibited, their
ovaries shrink further, often irreversibly
(reviewed by Wheeler 1986). If subordination can
impose long-term physiological inhibition and

ovary regression, an I. purpureus queen could
reduce competition over sexual production by
dominance while there is still opportunity to do
so. In spite of the efforts of the dominant gyne
who would be queen, the subordinate is released
from inhibition before sexuals are reared.
However, if either queen did not compete
behaviorally in the early phases, she would risk
permanent inhibition by the other. Though costly,
dominance interactions constitute an
evolutionarily stable strategy (Maynard Smith
1976), while peaceful coexistence would easily be
invaded by a dominance strategy.

The production and proximity of eggs
apparently involve two interrelated effects:
stimulation of the motivational state of dominance
and/or simulation of the well-developed ovaries
and hormonal states of high fertility. The data
suggest that subordinate queen Y actively
oviposited more frequently during encounters,
thereby obtaining a temporary dominance advantage,
leading on occasion to physical attacks.
Withholding eggs to lay during bouts, i.e. giving
the false appearance of a higher oviposition rate
and higher rank, could intimidate a rival and,
perhaps, increase self-motivation. In other, more
loaded, words both "deception" and "self-
deception" in the sense of Trivers (1985) could
result from the imitation of a more dominant
physiology. This phenomenon may not be unique to
I. purpureus. Wilson (1975) found that, while or
just before laying an egg, Leptothorax
curvispinosus Mayr workers enslaved in colonies of
L. duloticus L.G. Wesson sometimes attacked the
slave-maker queen.

A dominant I. purpureus queen usually stands
atop the eggs, where she is more likely to win,
and antennates longer during bouts, and an
increase in oviposition is observed in the
subordinate when she is in sole possession of the
egg pile. During phase II, when dominance was
destabilized, both queens gained an advantage from
being on the egg pile during a bout. Dominating
queens of other ants also tend to position
themselves on or near piles of eggs (Wilson 1974,
1975; Traniello 1982; Tschinkel and Howard 1983;
Evesham 1984). Possibly an ovarian substance,

generally indicative of fertility, that coats the
eggs is involved in both the stimulatory effect of
the proximity of eggs and that of actual
oviposition.
     The egg-laying and egg pile effects, the
tendency of the currently dominant or resident
queen to ignore the other's initiation, not to
retreat after losing and to be a "sore loser"
(contesting bouts she terminates with prolonged
antennation), the temporary improvement in status
by having won 1 of the last 5 bouts, and the
alterations in dominance during reunions in the
incomplete separation phase, all suggest that
motivation plays an extremely important role in
queen-queen interactions. Each queen's
experiences in prior encounters feed back into her
current "expectation" of winning, with both short-
and long-term influences, and she behaves
accordingly. A previously subordinated queen is
less motivated in subsequent encounters, while a
previously dominant queen is more so. Immediate
motivational enhancements, as by laying an egg or
having won a recent bout, are superimposed on (and
can reinforce or subvert) the more general
influence of habitual high or low status during a
given phase. All of these correlations may be
linked to a common physiological mechanism of
motivation. Over the longer time frame, the
mechanism probably involves juvenile hormone,
known to affect both oviposition rates and
dominance aggressiveness in bees and wasps (Breed
1982; Röseler et al. 1980, 1984).
     Queen-queen interactions were largely
reversed in phase II, when the previously
subordinate queen Y became resident and locally
dominant. By contrast, queen-worker interactions
indicated no significant bias based on dominance
status before separation, from the point of view
of the workers. (The subordinate or intruding
queen did antennate workers more frequently.) A
preference for the resident developed only after
dominant queen B moved out, and workers remained
tolerant of both. Thus the contribution to queen
ranking by workers, which attack or drive out
less-preferred queens of some species (Bartz and
Hölldobler 1982; Fletcher and Blum 1983) does not
occur in I. purpureus. Experiments with marked

workers once the queens had separated revealed no
fidelity of workers to particular queens
(Hölldobler and Carlin 1985). This also indicates
that early dominance interactions do not represent
each queen's attempt to lay the majority of worker
eggs, in order to produce a large daughter cohort
that will preferentially aid her. Marking results
did reveal that both workers and queens tend to
remain in particular regions within the nest, so
that queens may be in the company of the same
workers for extended periods. (Halliday [1983]
also demonstrated worker fidelity to particular
entrances of a large field colony of _I.
purpureus_.) Workers thus appear to be sensitive
to residence, but not dominance, status, and their
preferences could help maintain the separation of
cofoundresses in later phases.

Further investigation of additional foundress
groups is needed to address major unanswered
questions, such as why it was the dominant queen
who moved out of the founding chamber, and whether
she usually does so. Relatedness among
cofoundresses might also strongly affect their
interactions. Since many queens return to found
in the vicinity of what are probably their natal
nests, recognized and aided by the workers
(Hölldobler and Carlin 1985), frequent
associations of sister queens are not unlikely.
Allozyme analyses of within-colony genetic
heterogeneity (Halliday 1979, 1983) and the
readiness of queens collected from different
locales to cofound show that kinship is not
required, however.

To examine interindividual variability in
colonies whose workers may live for over a year
and queens for a decade, detailed long-term
observations of many identified individuals will
be necessary. In due course such studies may
reveal that hymenopteran colony politics is at
least as intriguing as that of chimpanzees.

SUMMARY

Pairs of cofounding queens in pleometrotic
associations of the Australian meat ant,
Iridomyrmex purpureus (Smith), engage in striking

antagonistic behavior, in the form of ritualized antennation bouts. Detailed observations of antennation behavior among a pair of individually-marked queens indicate that a reproductive rank order results, in which the dominant queen inhibits egg-laying by the subordinate. Eventually the queens become intolerant of each other's presence and permanently separate within the nest, after which they lay eggs at equal rates. Since dominance interactions have long since ended by the time the colony begins to produce reproductives, the selective advantage of this behavior early in coloy development is not obvious; possibly the dominant queen attempts to reduce later competition over alate production by permanently subordinating her cofoundress while they are still together.

ACKNOWLEDGMENTS

I am grateful to Bert Hölldobler for providing the problem, the ants, and additional field and laboratory data, and for reading the manuscript. David Gladstein gave indispensable assistance with data analysis. Research was supported by NSF grants BNS 80-02163 and BNS 82-19060 to Bert Hölldobler; the author's participation in the E.S.A./I.U.S.S.I. symposium on interindividual variability was supported by an award from the Department of Organismic and Evolutionary Biology, Harvard University.

REFERENCES

Bartz, S.H., and Hölldobler, B. 1982. Colony founding in Myrmecocystus mimicus Wheeler (Hymenoptera: Formicidae) and the evolution of foundress associations. Behavioral Ecology and Sociobiology 10:137-147.
Bhatkar, A., and Whitcomb, W.H. 1970. Artificial diet for rearing various species of ants. Florida Entomologist 53:229-232.

Breed, M. 1982. Juvenile hormone and aggression in the honey bee. In The biology of social insects. ed. M.D. Breed, C.D. Michener, and H.E. Evans, 233-237. Boulder, Colorado: Westview Press.

Brothers, D., and Michener, C.D. 1974. Interactions in colonies of primitively social bees III. Ethometry of division of labor in Lasioglossum zephyrum (Hymenoptera: Halictidae). Journal of Comparative Physiology 90:129-168.

Cole, B.J. 1981. Dominance hierarchies in Leptothorax ants. Science 212:83-84.

Ettershank, G. 1971. Some aspects of the ecology and nest microclimatology of the meat ant, Iridomyrmex purpureus (Sm.). Proceedings of the Royal Society of Victoria 84:137-152.

Evesham, E.J.M. 1984. Queen distribution movements and interactions in a semi-natural nest of the ant (Myrmica rubra) L. Insectes Sociaux 31:5-19.

Fletcher, D.J.C., and Blum, M.S. 1983. Regulation of queen number by workers in colonies of social insects. Science 219:312-315.

Fletcher, D.J.C., and Ross, K.G. 1985. Regulation of reproduction in eusocial Hymenoptera. Annual Review of Entomology 30:319-344.

Fowler, H.G., and Roberts, R.B. 1983. Anomalous social dominance among queens of Camponotus ferrugineus (Hymenoptera: Formicidae). Journal of Natural History 17:185-187.

Franks, N.R., and Scovell, E. 1983. Dominance and reproductive success among slave-making worker ants. Nature 304:724-725.

Gadagkar, R. 1980. Dominance hierarchy and division of labor in the social wasp, Ropalidia marginata (Lep.) (Hymenoptera: Vespidae). Current Science 497:72-75.

Greaves, T., and Hughes, R.D. 1974. The population biology of the meat ant. Australian Journal of Zoology 13:329-351.

Halliday, R.B. 1979. Esterase variation at three loci in meat ants. Journal of Heredity 70:57-61.

Halliday, R.B. 1983. Social organization of meat ants, Iridomyrmex purpureus, analysed by gel electrophoresis of enzymes. Insectes Sociaux 30:45-56.

Hölldobler, B. 1962. Zur Frage der Oligogynie bei Camponotus ligniperda Latr. und Camponotus herculeanus L. (Hymenoptera: Formicidae). Zeitschrift für Angewandte Entomologie 49:337-352.

Hölldobler, B., and Carlin, N.F. 1985. Colony founding, queen dominance and oligogyny in the Australian meat ant Iridomyrmex purpureus. Behavioral Ecology and Sociobiology 18:45-58.

Hölldobler, B., and Taylor, R.W. 1983. A behavioral study of the primitive ant Nothomyrmecia macrops Clark. Insectes Sociaux 30:384-401.

Hölldobler, B., and Wilson, E.O. 1977. The number of queens: an important trait in ant evolution. Naturwissenschaften 64:8-15.

Jeanne, R.L. 1972. Social biology of the Neotropical wasp Mischocyttarus drewseni. Bulletin of the Museum of Comparative Zoology, Harvard University 144:63-150.

Jeanne, R.L. 1980. Evolution of social behavior in the Vespidae. Annual Review of Entomology 25:371-396.

Lenoir, A. 1984. Brood-colony recognition in Cataglyphis cursor worker ants (Hymenoptera: Formicidae). Animal Behaviour 32:942-944.

Maynard Smith, J. 1976. Evolution and the theory of games. American Scientist 64:41-45.

Mintzer, A. 1979. Colony foundation and pleometrosis in Camponotus (Hymenoptera: Formicidae). Pan-Pacific Entomologist 55:81-89.

Pardi, L. 1948. Dominance order in Polistes wasps. Physiological Zoology 21:1-13.

Rissing, S., and Pollock, G. 1986. Social interaction among pleometrotic queens of Veromessor pergandei (Hymenoptera: Formicidae) during colony foundation. Animal Behaviour 34:226-233.

176

Röseler, P. 1965. Beobachtungen uber die Verhaltensweisen in Kunstlich erzielten polygynen Hummelvölkern. Insectes Sociaux 12:105-116.

Röseler, P., Röseler, I., and Strambi, A. 1980. The activity of corpora allata in dominant and subordinated females of the wasp Polistes gallicus. Insectes Sociaux 27:97-107.

Röseler, P., Röseler, I., Strambi, A., and Augier, R. 1984. Influence of insect hormones on the establishment of dominance hierarchies among foundresses of the paper wasp Polistes gallicus. Behavioral Ecology and Sociobiology 15:133-142.

Schmitz, H. 1911. Über die selbstandige Koloniegründung und die Folgen Künstlicher Pleometrose bei Camponotus ligniperda Ltr. Deutsche Entomologische National-Bibliotek 21:166-168.

Stumper, R. 1962. Sur un effet de groupe chez les femelles de Camponotus vagus (Scopoli). Insectes Sociaux 9:329-333.

Taki, A. 1976. Colony founding of Messor aciculatum (Fr. Smith) (Hymenoptera: Formicidae) by single and grouped queens. Physiology and Ecology, Japan 17:502-512.

Traniello, J.F.A. 1982. Population structure and social organization in the primitive ant, Amblyopone pallipes (Hymenoptera: Formicidae). Psyche 89:65-80.

Trivers, R. 1985. Social evolution. Menlo Park, California: Benjamin/Cummings Publishing Co.

Tschinkel, W.R., and Howard, D.F. 1983. Colony founding by pleometrosis in the fire ant Solenopsis invicta. Behavioral Ecology and Sociobiology 12:103:113.

Waloff, N. 1957. The effect of the number of queens of the ant Lasius flavus (Fab.) (Hymenoptera: Formicidae) on their survival and on the rate of development of the first brood. Insectes Sociaux 4:391-408.

West, M.J. 1967. Foundress associations in polistine wasps: dominance hierarchies in the evolution of social behavior. Science 157:1584-1585.

West-Eberhard, M.J. 1977. The establishment of reproductive dominance in social wasp colonies. In Proceedings of the 8th International Congress of the International Union for the Study of Social Insects, Wageningen, ed. H.H.W. Velthuis and J.T. Wiebes, 223-227. Wageningen, The Netherlands: Center for Agricultural Publishing and Documentation.

West-Eberhard, M.J. 1978. Temporary queens in Metapolybia wasps: Nonreproductive helpers without altruism? Science 200:441-443.

Wheeler, D.E. 1986. Developmental and physiological determinants of caste in social Hymenoptera: Evolutionary implications. American Naturalist 128:13-34.

Wilson, E.O. 1971. The insect societies. Cambridge, Massachusetts: Belknap Press of Harvard University Press.

Wilson, E.O. 1974. Aversive behavior and competition within colonies of the ant Leptothorax curvispinosus. Annals of the Entomological Society of America 67:777-780.

Wilson, E.O. 1975. Leptothorax duloticus and the beginnings of slavery in ants. Evolution 29:108-119.

7

Pleometrosis and Polygyny in Ants

Steven W. Rissing and Gregory B. Pollock

While most ant colonies are founded by a
single queen, a condition known as
"haplometrosis," colony foundation by groups of
queens, or "pleometrosis," occurs in a number of
species (see below). Groups of co-founding ant
queens permit a comparison to groups of cofounding
wasp queens and provide an opportunity to examine
behavioral variation among members of a group of
social insects in the absence of kin selection.
Here we review some of the available literature on
multiple queens in ant nests either at the
founding stage or later in colony development. We
address some of the selective forces behind
pleometrosis and their effect on the behavior of
co-founding queens. The fate of pleometrotic
queens is considered with special emphasis on the
occurrence and distribution of multiple queens in
adult colonies.

THE OCCURRENCE OF PLEOMETROSIS

While Wheeler (1910) argued that colony
foundation in most ants is haplometrotic, he also
provided some of the first data exceptional to
this rule (Wheeler 1917). In addition to his
early note on the honey ant Myrmecocystus mimicus
Wheeler, at least 12 other ant species generally
found colonies with groups of queens (Table
7.1). Some observations of pleometrosis, for
example Pogonomyrmex rugosus Emery and Pheidole
dentata Mayr (Table 7.1) are likely "purely

Table 7.1
Pleometrosis in ants[a]

Species	No. Nests Counted	Percent with 1 Queen	Percent with > 1 Queen	Reference
Rhytidoponera confusa	40	95	5	Ward 1981
Pogonomyrmex rugosus	70	94	6	Pollock & Rissing 1985
P. rugosus	"hundreds"[b]	100	0	Hölldobler 1976a
Pheidole bicarinata	125	98	2	Wheeler 1982
Veromessor pergandei	132	32	68	Pollock & Rissing 1985
Messor aciculatum	[b]	8	92	Namigai and Onoyama cited in Taki 1976
Solenopsis invicta	115	27	73	Tschinkel & Howard 1983
Iridomyrmex purpureus	72	90	10	Hölldobler & Carlin 1985
Acromyrmex versicolor	64	44	56	Rissing et al. 1986
Atta texana	97	97	3	Echols 1966
Camponotus modoc	204	40	60	G. Nielsen, pers. comm.
C. vicinus	11	91	9	G. Nielsen, pers. comm.
Lasius flavus	12	17	83	Waloff 1957
L. flavus	14	64	36	Talbot 1965
Myrmecocystus mimicus	20	ca. 50	ca. 50	Wheeler 1917
M. mimicus	ca. 387	28	72	Bartz & Hölldobler 1982

[a] Only studies of 10 or more colonies are reported.
[b] Cannot be determined given method of data presentation.

fortuitous" as argued by Baroni-Urbani
(1968a:269). For other species, however, colony
foundation by groups is quite common (Table
7.1). Pleometrotic queens lose less weight during
claustral production of their first brood, produce
more brood/workers and/or produce workers faster
than do haplometrotic queens of the same species
(Waloff 1957; Stumper 1962; Markin et al. 1972;
Hage and Hanna 1975; Taki 1976; Terron 1977;
Mintzer 1979; Bartz and Hölldobler 1982; Tschinkel
and Howard 1983; Rissing and Pollock 1987).

The likely advantage of rapid production of a
larger worker force by pleometrotic colonies
appears to exist in the phenomenon of brood
raiding between conspecific colonies found in at
least three highly pleometrotic ant species: M.
minicus (Bartz and Hölldobler 1982), the imported
fire ant Solenopsis invicta Buren (Tschinkel and
Howard 1983) and the desert seed-harvester
Veromessor pergandei (Mayr) (Rissing and Pollock
1987). These species also share a distinct
tendency to clump starting nests (M. minicus and
S. invicta: above references; V. pergandei:
Pollock and Rissing 1985) and to have sharply
defined and defended territories around adult
nests (M. minicus: Hölldobler 1976b, 1981; S.
invicta: Wilson et al. 1971; V. pergandei: Went
et al. 1972; Wheeler and Rissing 1975a; Byron et
al. 1980; Ryti and Case 1984). By rapidly
producing a larger raiding force, pleometrotically
founded colonies enjoy an advantage in early
territorial encounters. Without clumped starting
nests such an advantage does not exist; in such
cases pleometrosis, if present, must be due to
oligogyny (see below for an example of
pleometrosis independent of intraspecies
territoriality).

That pleometrosis can grant a territorial
advantage has been demonstrated recently in V.
pergandei (Rissing and Pollock 1987). We started
haplometrotic and pleometrotic (3 or 4 queens)
colonies of V. pergandei in soil-filled glass
bottles which were inserted through the floor of
plastic shoe boxes (one of each type of nest per
box) in the laboratory. Brood raiding between the
bottles (frequently simultaneous and reciprocal)
occurred soon after both nests opened

and began to forage. Twenty weeks after colony
initiation, all bottles were emptied and contents
censused. In 16 of the 19 boxes where both the
haplometrotic and pleometrotic bottles opened,
only the pleometrotic bottle survived; the
haplometrotic bottle was the sole survivor in only
one of the remaining boxes.

The ultimate cause for clumping of founding
nests of pleometrotic species seems associated
with abiotic constraints. Veromessor pergandei,
which flourishes in some of the driest habitats of
North America (Creighton 1950; Wheeler and Rissing
1975a, b) clumps its nests in ravine bottoms
(Rissing, unpublished data), which may provide
better access to ground moisture. The notoriously
deep nests of this species (Tevis 1958) may make
access to ground water more likely and obviate
problems associated with nest location in flash
flood-prone ravines. Solenopsis invicta queens
appear to clump nests on higher microsites when
available (Tschinkel and Howard 1983), a possible
adaptation to flooding in the native, river bottom
habitat of this species.

In the highly pleometrotic desert leaf-cutter
ant Acromyrmex versicolor (Pergande) clumping of
founding nests is clearly associated with
avoidance of potentially lethal temperatures at
all places in its habitat save for shaded areas
around the bases of large trees. Founding nests
are located almost exclusively here (Rissing et
al. 1986). This may be associated with the
invasion of the dry desert habitat by this and the
ecologically similar (and pleometrotic: Moser
1967, Mintzer and Vinson 1985) Atta texana
(Buckley) in the otherwise largely tropical group
of leaf-cutter ants (Creighton 1950; Weber
1972). Adult territoriality in the leaf-cutter
group (Gamboa 1974), when coupled with the
exclusive association of a single adult nest with
a large tree regardless of starting nest density,
suggest that immature colonies experience severe
territorial competition. We predict that brood
raiding must exist among starting colonies as the
proximate selective force behind pleometrosis in
Acromyrmex versicolor.

Relatedness of Co-foundresses

Wheeler (1910:190) also supported the notion
that when group founding did occur, co-foundresses
were "a couple of females from the same maternal
nest (who) may meet after their marriage flight
and together start a colony." This argument has
provided a convenient "ad hoc modification" of
Hamiltonian (1964) kin selection and the apparent
need of a single foundress (singly inseminated) to
maintain high degrees of relatedness among sterile
workers and reproductives of a colony. The notion
that multiple ant foundresses are closely related
(usually sisters) gained much support from studies
of primitively eusocial wasps. Queens of Polistes
spp., for example, start colonies only with
maternal nestmates when given a choice in
laboratory tests (Pfennig et al. 1983 and included
references). Pleometrotic ant queens given the
same sort of choice in the laboratory, however,
display no such preferences. We have collected V.
pergandei queens from locations up to 1-6 km
apart. These queens were either under rocks or
other debris, frequently in groups, or had
recently begun to excavate a nest (again,
frequently in a group). By marking queens with
respect to collection locale (controlling for
paint odor by using different sequences of the
same color or no paint at all), we were able to
show that V. pergandei queens not only tolerate
any conspecific queen but will co-found without
any respect to initial collection locale (and
hence without respect to a specified degree of
relatedness necessary for the evolution of
altruism through kin selection; Rissing and
Pollock [1986]). Recent work with Acromyrmex
versicolor demonstrated co-founding again without
respect to collection locale in this species
(Rissing et al. 1986). Founding groups of M.
mimicus (Bartz and Holldöbler 1982), Atta texana
(Mintzer and Vinson 1985), Iridomyrmex purpureus
(Smith) (Hölldobler and Carlin 1985) and Messor
aciculatum (Fr. Smith) (Taki 1976; Onoyama 1981)
also associate without respect to collection
locale. Electrophoretic evidence indicates that
multiple founding queens of S. invicta are no more
related to each other than are queens chosen

randomly from the population (Ross and Fletcher
1985) Taken together, these observations suggest
that the selective dynamics behind queen
associations are different between co-founding
ants and their wasp counterparts.

Behavior of Pleometrotic Ant Queens

Some of the most impressive examples of
interindividual variation in behavior among social
insects occur among pleometrotic wasps where a
distinct dominance hierarchy forms rapidly; by the
time the first workers eclose, a dominant, alpha
queen is in control of the colony with other co-
foundresses subordinate to her (Pardi 1948; West
1967; West Eberhard 1969; Jeanne 1972; Litte 1979;
Noonan 1981). Pleometrotic ant species appear
different from the above "wasp model." While ants
have not been studied as well as wasps, no
dominance hierarchies appear to form during colony
foundation. Further, in lieu of subordination,
multiple queens are simply killed following
initial worker production.
Evidence that pleometrotic ant queens produce
more brood and workers (references cited above)
suggests that all queens play an active
reproductive role during colony foundation. All
V. pergandei queens in laboratory observation "ant
farms" share in the task of nest excavation and,
most importantly, share equally in production and
care of the first brood (Rissing and Pollock 1986,
1987). "Functional monogyny" more closely
resembling interactions among multiple wasp queens
occurs among adult colonies of several ant species
(see below), but has yet to be documented during
the founding stage of any ant colony. Maintenance
of separate and distinct brood piles, not observed
in V. pergandei in our laboratory "ant farms," has
been reported in single colonies of Camponotus
ferrugineus (Fabricius) (Fowler and Roberts 1983)
and Iridomyrmex purpureus (Hölldobler and Carlin
1985) (see also discussion of Lasius flavus Mayr
below).
When the first V. pergandei workers are
produced and begin foraging, queens begin to
fight, frequently to death (Rissing and Pollock

1987). Worker involvement in such fights is secondary and directed toward recent participants in fights who are often visibly damaged or even dead. Other observations of worker elimination of initial co-foundresses (Baroni-Urbani 1968a; Hölldobler and Wilson 1977; Ward 1981; Bartz and Hölldobler 1982) may represent similar behavior following queen fights. If workers are going to expel all but one of the original foundresses, selection should shape the behavior of potential 'victim' queen(s) to be selected by the workers for survival. This can be accomplished by rendering rival queens less acceptable to workers: that is, injuring them through queen fights prior to the eclosion of the main worker force. Behavior similar to worker ejection of "excess" co-foundresses can be elicited in laboratory colonies by mixing independent, adult colonies or portions thereof (Baroni-Urbani and Soulié 1962; Soulié 1964; Hölldobler and Wilson 1977; Provost 1979) and may also follow some form of queen conflict.

The shift in behavior of pleometrotic ant queens during eclosion of the first workers and opening of the nest may reflect at least in part the claustral nature of ant colony foundation. While claustral, any energy expended to dominate co-foundresses is lost from production of the first worker force--the apparent selective 'goal' of pleometrosis in the first place. At this point in colony development, competition to gain control of the nest, or more appropriately, its worker force, can only lessen the value of that resource (Rissing and Pollock 1986; Pollock, in review). Colony foundation under these circumstances is essentially a mutualistic process (Bartz and Hölldobler 1982; Ross and Fletcher 1985; Rissing and Pollock 1986). This situation should generally change as workers eclose and begin to forage: given reasonable success at foraging (and/or brood raiding), the value of the resource (the worker force) should not be affected by queen conflict. It is precisely at this point that V. pergandei queens begin to fight (Rissing and Pollock 1987). It is also at this point that a starting, previously claustral, colony is most ergonomically stressed (Hölldobler and Wilson

1977; Oster and Wilson 1978). The entire colony
energy "intake" has consisted only of the fixed
fat reserves of the founding queens; not
surprisingly, queens have lost over 50 percent of
their starting mass by the time the nest opens
(Rissing and Pollock 1987). Queens essentially
compete for food harvested by the initial workers;
elimination of just one queen should remove a
major 'drain' on the energy intake of a starting
colony at this early stage.

An intriguing variation on the above theme
exists in Acromyrmex versicolor, which, unlike
most other higher ants, is not claustral:
starting queens forage for plant material prior to
first worker eclosion (Weber 1972; Gamboa 1974).
Given abundant pleometrosis in this species, an
opportunity exists to test the extent of mutualism
among founding queens (Rissing et al. 1986).

FATE OF PLEOMETROTIC QUEENS: OCCURRENCE OF
MULTIPLE QUEENS IN ADULT COLONIES

While our observations with laboratory
colonies of Veromessor pergandei indicate that
most colonies reduce to monogyny following opening
of the nest, this is certainly not always the
case. In the pleometrotic nest bottles discussed
above an average of 1.85 (± 1.01) queens were
still alive about 10 weeks following opening of
the nest; due to excavation of the nest bottles at
this time, queen mortality data following this
point are inconclusive. V. pergandei colonies at
least 12 weeks old with as many as 140 workers
have been excavated in the field with as many as
nine clumped queens, all capable of laying eggs
(Wheeler and Rissing 1975b). Such observations,
coupled with the high incidence of multiple queens
in adult ant colonies (see below) suggests that
immediate death may not always be the fate of
multiple co-foundresses in ants. We have
tabulated data from reports of colony counts for a
minimum of ten colonies for a given species. The
arbitrary ten colony minimum is meant to ensure a
reasonable estimate of population patterns for a
species before it is considered here. The effect
of this minimum is to generate a conservative

estimate of polygyny among ants since notes
regarding single observations of unclear
significance (for example, for Atta colombica
Santschi [Martin et al. 1967]) and observations of
species that have very large colonies but are
nonetheless polygynous (for example, Formica
exsectoides Forel [Cory and Haviland 1938]) tend
to be excluded. Some form of polygyny occurs in
72 percent of the 60 species surveyed (Table
7.2). This list is by no means exhaustive
(especially since investigators tend not to
publish long lists of colony censuses when species
are apparently "normal"--that is, monogynic).
Several hypotheses have addressed the presence of
multiple queens in adult ant colonies. These are
discussed below.

Oligogyny

An alternative to queen aggression and death
at the end of claustrality is the establishment of
distinct worker forces and territories by initial
co-foundresses within a single nest, that is,
"oligogyny" (Hölldobler 1962). However, if
selection for pleometrosis is largely through an
enhanced worker force, oligogyny could cancel this
advantage. Pleometrotic colonies produce fewer
workers/queen than haplometrotic colonies (Waloff
1957; Wilson 1974; Taki 1976; Terron 1977;
Berkelhamer 1980). If queens divide the pre-
worker brood and establish independent units, a
queen should, on average, have fewer workers than
a haplometrotic counterpart. Co-founding queens
of Lasius flavus become oligogynous prior to first
worker eclosion (Waloff 1957; Talbot 1965). If
workers of this species exhibit site fidelity to
their area of eclosion (and hence to the nearby
queen) oligogyny effectively reduces the worker
force of a given queen. Such worker site fidelity
occurs in the polygynous/oligogynous Pseudomyrmex
venefica Wheeler (Beulig and Janzen 1969; Janzen
1973) and Iridomyrmex purpureus (Halliday 1983;
Hölldobler and Carlin 1985). Similarly, Myrmica
rubra L. workers prefer to remain with a single
queen in their polygynous colonies; indeed, "it is
possible workers recall a place of origin"

Table 7.2
Monogyny and polygyny in adult ant nests[a]

Species	Population Structure[b]	No. Nests Counted	Percent with 0 Queens	Percent with 1 Queen	Percent with >1 Queen	Reference
Myrmeciinae						
Myrmecia dispar	M	35	9	89	3	Gray 1971
M. pilosula	M	14	c	36	64	Craig & Crozier 1979
M. varians	M	16	0	94	6	Gray 1974
Aneuretinae						
Aneuretus simoni	P	26	66	19	8	Jayasuriya & Traniello 1986
Pseudomyrmecinae						
Pseudomyrmex belti	M	100[d]	c	100	0	Janzen 1973
P. ferruginea	M	100[d]	c	100	0	Janzen 1973
P. nigrocincta	M	100[d]	0	100	0	Janzen 1973
P. venefica	U	12	0	0	100	Janzen 1973

Dolichoderinae

Dolichoderus plagiatus	M	12	33	58	8	Kannowski 1967
D. pustulatus	M	41	20	66	15	Kannowski 1967
D. quadri-punctatus	M/P?	287	77	23	0	Torossian 1960
Tapinoma sessile	M	10	c	c	0	Smith 1928
Iridomyrmex pruinosum[e]	?	27	-	81	19	Berkelhamer 1980
Conomyrma bicolor[e]	M	24	-	58	42	Berkelhamer 1980
C. insana[e]	M	34	-	100	0	Berkelhamer 1980
C. insana	U	27	c	c	19	Nickerson et al. 1975
C. flavocepta	M	34	71	29	0	Nickerson et al. 1975

Formicinae

Formica pallidefulva schaufussi incerta	M/U?	24	4	29	67	Talbot 1948
F. pallidefulva	M	24	0	100	0	Talbot 1948
F. obscuripes	U	28	0	0	100	Finnegan 1977
F. japonica	U	20	5	5	90	Higashi 1979
F. yessensis	U	49	16	10	73	Ito 1973
F. yessensis	U	10	30	0	70	Higashi 1983
F. transkau-casica[e]	M/P?	35	-	34	66[d]	Pamilo 1982

(continued)

Table 7.2 (cont.)

Species	Population Structure[b]	No. Nests Counted	Percent with 0 Queens	Percent with 1 Queen	Percent with >1 Queen	Reference
F. aquilonia	U	30	-	3	97[d]	Pamilo 1982
F. polyctena[a]	U	17	-	0	100	Pamilo 1982
Lasius sakagamii	U	108	20	18	62	Yamauchi et al. 1981, 1982
Prenolepis imparis	M	20	15	75	10	Talbot 1943
Camponotus ferrugineus	M	19	0	100	0	Pricer 1908
C. pennsyl-vanicus	M	57	4	93	4	Pricer 1908
C. texanus	M/P?	16	12	25	63	Creighton & Gregg 1954
C. werthi	M/P?	12	75	25	0	Skaife 1961
Ponerinae Amblyopone pallipes	U	31	39	29	32	Traniello 1982
Dorylinae Neivamyrmex carolinensis	M	22	0	55	45	Rettenmeyer & Watkins 1978

Species	Type					Reference
N. nigrescens	M	20[d]	0	100	0	Rettenmeyer & Watkins 1978
N. opacithorax	M	10[d]	0	100	0	Rettenmeyer & Watkins 1978
Myrmicinae						
Myrmica emeryana	M	36	8	78	14	Talbot 1945
M. lobicornis	M	15	13	7	80	Kannowski 1970
M. rubra[f]	P?	265	5	c	57	Elmes 1973
M. rubra[f]	M	24	0	0	100	Elmes 1974a
M. ruginodis[f]	M	83	45	40	16	Elmes 1980
M. ruginodis (Woodland form)	M	20	5	65	30	Mizutani 1981
M. ruginodis (Riverside form)	U	60	0	0	100	Mizutani 1981
M. sabuleti[f]	M	33	3	33	64	Elmes 1974a
M. scabrinodis	M	26	31	27	42	Elmes 1974a
M. schenki	M	29	45	38	17	Elmes 1980
M. sulcinodis	M	40	50	48	2	Elmes 1974b
Stenamma brevicorne	M	15	53	40	7	Talbot 1975
S. diecki	M	10	30	70	0	Letendre & Pilon 1972
S. diecki	M	15	27	60	13	Talbot 1975
Aphaenogaster rudis	M	10	30	70	0	Headley 1949
A. rudis	M	71	3	73	24	Talbot 1951
A. treatae	M	30	7	93	0	Talbot 1954
A. japonica	M/P?	13	54	46	0	Mizutani & Imamura 1980

(continued)

Table 7.2 (cont.)

Species	Population Stucture[b]	No. Nests Counted	Percent with 0 Queens	Percent with 1 Queen	Percent with >1 Queen	Reference
Leptothorax ambiguus	M/P	1522	30	50	20	Alloway et al. 1982
L. curvispinosus	M/P	38	32	67	0	Headley 1943 (Alloway et al. 1982)
L. curvispinosus	M/P	525	28	66	6	Talbot 1957
L. curvispinosus	M/P	488	36	47	17	Alloway et al. 1982
L. longispinosus	P	84	26	64	10	Headley 1943 (Alloway et al. 1982)
L. longispinosus	P	2650	33	49	18	Alloway et al. 1982
L. longispinosus	P	178	28	43	29	Herbers 1984
L. muscorum	?	27	18	37	44	Buschinger 1979a
L. provancheri	M/P?	15	0	7	93	Buschinger et al. 1980
Formicoxenus hirticornis	M	12	8	50	42	Buschinger 1979b
Harpagoxenus canadensis	M	6	9	87	4	Buschinger & Alloway 1978
H. sublaevis	?	54	6	94	0	Buschinger 1966

Species						Reference
H. americanus	P	31	52	39	10	Wesson 1939
H. americanus	P	15	27	73	0	Headley 1943 (Buschinger & Alloway 1977)
Solenopsis geminata	M	11	c	c	73	Adams et al. 1976
S. invicta	M	35	c	c	17	Glancey et al. 1973
S. invicta	M	36	c	c	5	Markin et al. 1973
S. invicta (GA)	M	151	42	58	0	Fletcher et al. 1980
S. invicta (MS)	M	38	34	26	39	Fletcher et al. 1980
Pogonomyrmex montanus	M	80	16	84	0	MacKay 1981
P. occidentalis	M	33	18	82	0	Lavigne 1969
P. rugosus	M	20	20	80	0	Lavigne 1969
P. subnitidus	M	26	23	77	0	Lavigne 1969

[a] Only studies where 10 or more colonies have been examined are included here.
[b] M = multicolonial (see text), U = unicolonial (see text), P = polydomous, i.e., a distinct colony occupies two or more adjacent nest sites.
[c] Value cannot be determined given method of data presentation.
[d] Minimum estimate.
[e] Electrophoretic study.
[f] "Macrogyna" colonies only.

(Evesham 1984:153). Faster production of workers, the other major advantage of pleometrosis, should remain unaffected under primary oligogyny. The natural history of L. flavus suggests that this may be the primary advantage of pleometrosis in this species. Colonies farm aphids and seem able to live off them completely (Pontin 1978; Brian 1983). When workers farm underground rather than forage above ground, fewer of them seem necessary for the survival of a starting colony. Rapid production of workers should reduce demand on a queen's energy reserves, increasing her chances of surviving past claustrality.

Baroni-Urbani (1968a:272) has argued that "the majority of ant colonies in nature that at first glance appear polygynous should be interpreted as oligogynous." Most authors fail to note if multiple queens are grouped or isolated when found; nonetheless, queens have been reported grouped in Pseudomyrmex venefica (Janzen 1973); Formica japonica (Higashi 1979); Tapinoma sessile (Say) (Smith 1928); Conomyrma insana (Buckley) (Nickerson et al. 1975); Leptothorax ambiguus Emery, L. curvispinosus Mayr, and L. longispinosus Roger (Alloway et al. 1982; Herbers, 1986), Atta texana (Moser and Lewis 1981 and included references), Myrmica rubra (Evesham 1985), and Veromessor pergandei (Wheeler and Rissing 1975b). Such truly polygynous associations likely exhibit significantly different reasons for polygyny than oligogynous associations.

Functional Monogyny

The second hypothesis for the occurrence of multiple queens in adult colonies suggests that only a single queen in fact is reproductively competent. The best documented case of such "functional monogyny" occurs in monogynous populations of S. invicta. Tschinkel and Howard (1978) removed the single physogastric, apparently egg-laying queen from each of 114 S. invicta field nests; upon subsequent examination of 56 of these, replacement queens were found in 34. Seven of the replacement queens were fertile; the other 26 were uninseminated (presence of sperm determined by

type of brood produced; one of the 34 nests was
not accounted for). Only 2 of 114 nests observed
ever had more than one physogastric queen at one
time (and in at least one of these two cases, all
three queens were uninseminated). Further,
younger colonies (as determined by mound size)
were more likely to have replacement queens.
Consequently, Tschinkel and Howard (1978) argue
that in monogynous populations of S. invicta,
colonies are founded pleometrotically with a
single queen soon taking over all reproductive
function. Replacement queens (inseminated or not)
are thought to be original pleometrotic
foundresses subsequently released from
reproductive inhibition. Since associating queens
are unrelated (discussed above), this would imply
that reproductively inhibited queens exist for the
persistence of the colony. Further, it assumes
that auxiliary queens do not represent a
significant energetic cost for the functional
queen. Markin et al. (1972:1055) report that
approximately 50 percent of S. invicta colonies
are founded pleometrically, but "by the time
workers had emerged and were foraging on the
surface, only 1 queen remained in the nests
examined." This suggests that co-foundresses are
normally not retained as auxiliaries but rather
eliminated, consistent with Tschinkel and Howard's
(1978) observation that younger colonies were most
likely to have 'replacement' queens.
        The guest ant, Formicoxenus nitidulus (Nyl.),
has true queens, true workers (ergatomorphs), and
intermorphs between these two; all can contain
sperm although this is uncommon in the
ergatomorphs (Buschinger and Winter 1976). While
19 of 34 colonies examined had more than one
inseminated queen or intermorph, each colony
contained only a single egg-laying female
(Buschinger and Winter 1976). A similar situation
occurs in the guest ants Leptothorax provancheri
Emery (Buschinger et al. 1980) and Leptothorax
gredleri Mayr (Buschinger 1968). Functional
monogyny has also been reported for a single
colony of Myrmecina graminicola Latr. (Baroni-
Urbani 1968b), Harpagoxenus canadensis M.R. Smith
(Buschinger and Alloway 1978), and the
degenerative slave-maker Epimyrma kraussei Emery

(Buschinger and Winter 1983).

Two other cases of functional monogyny are less conclusive. While the guest ant Formicoxenus hirticornis (Emery) appears very similar to F. nitidulus as described above, 2 of 12 colonies examined had more than one functional egg layer (Buschinger 1979b). While Buschinger attributed this condition to a mixing of nests after (or in the process of) collecting, it is possible colonies are not always functionally monogynous. Similarly, in arguing that Myrmica sulcinodis Nyl. is monogynous, Elmes (1974b) dismisses a colony with 11 "old" queens by observing that 2 of 3 of these old queens were uninseminated when dissected. Functional mongyny in a species may not be absolute. For example, Buschinger (1979a) discusses queen number in colonies of Leptothorax muscorum (Nyl.); while a colony with 11 dealate queens contained only a single inseminated, fertile queen, two other colonies each contained two inseminated, egg-laying queens.

'Functional monogyny' in S. invicta (as discussed above) exhibits geographical variation. Glancey et al. (1973) found multiple dealate females in 6 of 35 colonies excavated. When isolated in laboratory colonies with workers, all of the multiple queens laid eggs within 4-11 days. Noting that this could be sufficient time to permit loss of reproductive inhibition and development of replacement queens (as in Tschinkel and Howard 1978), Fletcher et al. (1980) used direct observation of egg laying as well as ovary dissections to establish that 87 percent of 476 queens from 15 colonies were functionally polygynous. These functionally polygynous colonies were found at Hurley, Mississippi, while only monogynous colonies were found between Madison and Monticello, Georgia (Fletcher et al. 1980). The functionally monogynous colonies of this species studied by Tschinkel and Howard (discussed above) were from northern Florida (Fletcher et al. 1980). Variation in queen number in this species appears to be at least partially geographical and to have occurred following introduction into North America (Fletcher et al. 1980; Fletcher 1983; Greenberg et al. 1985; Ross and Fletcher 1985).

Functional Polygyny

Most of the studies upon which Table 7.2 is based were not performed to address the existence of functional monogyny or, for that matter, "functional polygyny," that is, the existence of more than one functioning egg-layer at a time in a colony. Nonetheless, accumulating evidence suggests that functional polygyny occurs among ants and is probably more common than functional monogyny. Evidence to support this comes either from electrophoretic studies examining relatedness among colony members or, as in the functional monogyny studies discussed above, direct observation/dissection of queens.

Electrophoretic evidence indicating functional polygyny exists for Myrmecia pilosula (F. Smith) (Craig and Crozier 1979), Conomyrma bicolor (Wheeler) and Iridomyrmex pruinosum (Roger) (Berkelhamer 1980, 1984) and several Formica spp. (Pamilo and Varvio-Aho 1979; Pamilo 1981, 1983; Pamilo and Rosengren 1984). An extensive series of dissections indicates that Leptothorax ambiguus, L. curvispinosus and L. longispinosus are commonly functionally polygynous (Alloway et al. 1982; see also Herbers 1984) confirming Wilson's (1974) observations of multiple egg layers in L. curvispinosus. Dissection has also revealed multiple, inseminated, egg laying queens in Formica japonica (Higashi 1979) and Formica integra (Kloft et al. 1973). Incorporation of oil-soluble dyes into the food of 14 queens of Solenopsis geminata indicates that all queens contributed to egg production in a laboratory study (Adams et al. 1976). Noting the enlarged appearance of the gaster of multiple females from nests of Conomyrma insana, Nickerson et al. (1975) argue this species is functionally polygynous in northern Florida. Observations of multiple-queen colonies of Amblyopone pallipes (Haldeman) and Aneuretus simoni (Emery) suggest these species are also functionally polygynous (Traniello 1982; Jayasuriya and Traniello 1986).

Indirect evidence indicating functional polygyny comes from a positive correlation between queen number and adult worker number in nests of some ant species (Table 7.3; see also Greenslade

Table 7.3
Correlation between number of queens and adult workers found in ant colonies[a]

Species	R^2	N	P	Reference
Myrmica emeryana	.00	33	NS	Talbot 1945
M. fracticornis	.02	13	NS	Kannowski 1970
M. rubra	.07	24	NS	Elmes 1974a
M. sabuleti	.26	32	.01	Elmes 1974a
M. scabrinodis	.27	18	.05	Elmes 1974a
Aphaenogaster rudis	.11	69	.01	Talbot 1951
Dolichoderus pustulatus	.33	33	.01	Kannowski 1967
Lasius sakagami (before flood)	.28	32	.01	Yamauchi et al. 1981, 1982
Formica pallidefulva	.12	23	NS	Talbot 1948
F. japonica	.07	14	NS	Higashi 1979

[a]Only those cases where there are data for at least 10 complete colonies are reported. Data for colonies without a queen are not used.

1971; Alloway 1980). Such a correlation is only inferential since it is possible that additional queens in a colony somehow aid the single, functional queen to raise more workers. Alternatively, failure to find a correlation (or finding only a weak one) does not necessarily prove functional monogyny, since all queens in a colony may reduce their production of eggs as queen number increases, as occurs among some pleometrotic foundresses (references cited above).

Studies of functional polygyny in ants have tended to be qualitative: a queen is either inseminated or she is not; her ovaries are functioning or they are not. Dissections to examine spermathecae and/or ovaries will not establish differential egg production among multiple egg-layers. Further complication arises when one considers that most progeny of an ant colony do not represent a direct increase in the fitness of the queen(s) that produce them. Asymmetry in laying eggs destined to become reproductive individuals occurs in the social wasp Polistes fuscatus (F.) (Metcalf and Whitt 1977) and is suggested to occur in the ant Formica sanguinea Latr. (Pamilo and Varvio-Aho 1979).

Secondary Polygyny

Multiple queens in ant colonies may be the result of adoption of additional queens by those colonies after they became adult (Baroni-Urbani 1968). Hölldobler and Wilson (1977) emphasize the coincidence between such "secondary polygyny" and unicolonial ant species, that is, those in which a single colony occupies many nests and, ultimately, the entire available habitat. Such habitats are frequently widely dispersed. Secondary polygyny is common in such unicolonial species (Table 7.2). Colonies frequently accept their own daughters back into the nest.

Five newly mated queens of the unicolonial acacia ant, Pseudomyrmex venefica, have been observed directly to re-enter the acacia thorn from which they emerged prior to mating (Janzen 1973). Alate females of the unicolonial Argentine ant, Iridomyrmex humilis (Mayr), commonly never even leave the nest; mating (apparently with brothers) occurs in the parent nest (Markin 1970). Adult colonies of the unicolonial _Lasius sakagamii_ Yamauchi & Hayashida retain their daughter queens (Yamauchi et al. 1981). Further, laboratory observations of _Amblyopone pallipes_ suggests that polygyny in this unicolonial species may be at least partly due to adoption of colony offspring following mating (Traniello 1982).

Some of the most dramatic examples of secondary polygyny are found in the diverse genus _Formica_ where functional polygyny and unicoloniality occur in populations of many species. In some populations of _F. sanguinea_, _F. exsecta_ Nyl., _F. polyctena_ Foerst and _F. pressilabris_ Nyl., multiple queens in a single nest are related and usually all lay worker eggs; adjacent nests are also related ("polydomy") (Pamilo and Varvio-Aho 1979; Pamilo 1981, 1983; Pamilo and Rosengren 1984). Similarly, newly mated queens of _F. yessensis_ Forel frequently enter neighboring nests and presumably their own as well (Higashi 1983). Polygynous populations of _F. transkaucasica_ Nasanov and _F. fusca_ L., however, do not appear to have related adjacent nests ("monodomy") even though the queens in a given nest are related (Pamilo 1982, 1983, 1984). Finally, in _F. aquilonia_ Yarrow, the multiple egg-laying queens in a given nest are no more related to each other than to the population as a whole (Pamilo 1982). In a number of _Formica_ species, distinctly different population structures can be found between closely related species or even among populations of the same species (reviewed in Pamilo and Rosengren 1984; see also above for _S. invicta_ and Halliday [1983] for _I. purpureus_). Pamilo and Rosengren (1984) suggest an evolutionary sequence from (presumably primitive) monogyny/monodomy to polygyny/polydomy, the latter step possibly accommodating decreasing relatedness among the multiple egg-layers within a given nest (see also Crozier 1979). This view is supported by apparently different "thresholds for nest splitting" in different populations of _F. sanguinea_ (Pamilo 1981; see also Markovsky 1963). Further, Pisarski (fidé Pamilo and Rosengren 1984) discusses a stage of

polygyny/monodomy in addition to the above two in
his studies of F. exsecta.

At least some of the Formica spp. can be
considered isolated habitat specialists in the
sense of Hölldobler and Wilson, for example, F.
polyctena, F. exsecta (Hölldobler and Wilson 1977)
and F. transkaucasica (Pamilo 1982). Habitat for
F. polyctena and F. aquilonia (at least in
Europe), however, is not patchily distributed
(Pamilo 1982). Alloway et al. (1982) draw on
Hölldobler and Wilson's (1977) observation that
unicolonial members must all have the same (or no)
nest odor, and use nest odor to establish polydomy
in Leptothorax ambiguus and L. longispinosus (and
possibly L. curvispinosus). It is difficult to
consider such polygynous species as "isolated
habitat specialists" since polydomous Leptothorax
colonies simply occupy more than one acorn.
Polygyny in Leptothorax ants seems a possible
example of polygyny occurring in species that
occupy short-lived nest sites, the second
environmental correlate associated with polygyny
discussed by Hölldobler and Wilson (1977).
Herbers (1986) argues that for this to be
effective, however, the multiple queens in a
Leptothorax nest should be physically separated
("to service most or all of the subunits"
following nest fragmentation [Hölldobler and
Wilson 1977:12]); multiple queens of L.
longispinosus, however, are clumped in both field
and laboratory nests. Further, Herbers (1986)
could find no correlation between degree of
polygyny and field measures of nest site fragility
in this species. Herbers (1986, 1987) argues that
competition for nest sites is a more likely
explanation for polygyny in Leptothorax; with
polygyny representing a backup of reproductive
daughters when nest sites are limited. Similarly,
limitation of nest sites may favor polygyny in
Formica (Pamilo 1981; Pamilo and Rosengren
1984). Such selective forces are essentially
similar to those forces causing nest clumping and
brood raiding in species such as A. versicolor, V.
pergandei, S. invicta and M. mimicus discussed
above.

Craig and Crozier (1979) extend the "isolated
habitat specialist" argument to Myrmecia pilosula,

a strictly multicolonial species, in an effort to
explain observed polygyny in this species. Such
efforts tend to so broaden the meaning of
Hölldobler and Wilson's strictly unicolonial
concept of "isolated habitat specialist" to render
it meaningless, since any species can ultimately
be regarded as a specialist to the habitat in
which it is found.

A set of marking observations suggests that
Myrmica rubra adopts fertilized daughters (Elmes
1973); electrophoretic evidence suggests that
queens in some, but not all polygynous colonies of
this species are closely related (Pearson 1982).
This species, however, may reproduce by budding
(Elmes 1973; Pearson 1982) and may more
appropriately be considered above as a unicolonial
species (although Elmes [1980] argues against the
possibility). Workers of Crematogaster
scutellaris Oliv. are reported to adopt any
conspecific, fertilized queen (Soulié 1962 [fidé
Brian 1965]).

Wilson (1963:251) notes that polygyny is more
common in certain rare ant taxa than in "commoner
members of the genus or nearest related taxa."
This should not be taken, however, as an
indication that polygyny occurs only in rare
species. This is certainly not the case for truly
unicolonial species who tend to monopolize entire
habitats (although see discussion by Crozier
[1979] regarding the evolution of polygyny under
unicoloniality). Numerous multicolonial species
are also polygynous and quite common. Myrmecia
pilosula (74 percent polygynous colonies; see
Table 7.1) is "very abundant in all eastern
states" of Australia (Clark 1951:204); Myrmica
rubra (57 percent polygyny) is abundant in
northern Europe (e.g. see numerous papers by
Brian); M. lobicornis (80 percent polygyny) is
"widely distributed" and "moderately abundant" in
North Dakota (Wheeler and Wheeler 1963:104). In
his comparison of ant communities in France and
North Africa, Bernard (1958:340) states "la
majorite des types dominants sont polygynes."
Similarly, Buschinger (1974) argues that
approximately half of all European ants are
polygynous.

The classification scheme of Hölldobler and
Wilson (1977) and Wilson's (1963) association of
polygyny and rareness, while incomplete, point to
the environment as a major force selecting for
polygyny in ant species. Even within a species,
queen number seems plastic and frequently varies
significantly with habitat (Table 7.4; see also
Bernard [1958] for variation in queen number in
Pheidole pallidula and above discussions of
Formica spp. and S. invicta). In almost every
case where sufficient data exist (i.e. at least 10
colonies from 2 sites), queen number varies
significantly across habitat. In addition to
species in Table 7.4, Conomyrma insana also varies
in queen number (and population-structure) in
different habitats (Table 7.2), as does Myrmica
sabuleti Meinert (Elmes and Wardlaw 1982). To
some extent these differences in queen number may
represent taxonomic difficulties in
differentiation of species (see possibility of C.
insana being a species complex [Snelling fidé
Berkelhamer 1984]; similarly the Formica
pallidefulva group treated as two species by
Talbot [1948] has now been grouped as a single
species [Smith 1979]). Queen number within a
given habitat also fluctuates through time in L.
longispinosus (Herbers 1984) and M. sabuleti
(Elmes and Wardlaw 1982).

FUNCTIONAL POLYGYNY AND KIN SELECTION

Equating polygyny largely with secondary
polygyny, mainly in unicolonial species, has
tended to focus attention away from the high
degree of facultative polygyny that occurs in many
multicolonial species (Table 7.2) and the
possibility that such polygyny indicates an
initial pleometrotic foundress association.
Pleometrotic foundation and polygynous adult
colonies, necessary but not sufficient conditions
to prove primary polygyny, occur in Myrmecia
pilosula (Wheeler 1933; Haskins and Haskins 1950;
Gray 1971), Leptothorax ambiguus and L.
longispinosus (Alloway et al. 1982), Tapinoma
simrothi Emery (Hage and Hanna 1975), Camponotus
texanus Wheeler (Creighton and Gregg 1954), and

Table 7.4
Queen number as a function of habitat[a]

Species	Location	Mean no. of queens (±SD)	N	t	P	Reference
Myrmica s. sabuleti	Hartland Moor Purbeck	1.40 (±.97) 3.91 (±3.15)	10 22	2.45	<.05	Elmes 1974a[b]
M. ruginodis	Woodland Riverside	2.1 (c) 11.2 (c)	c c		<.05	Mizutani 1981
Aphaenogaster rudis	Seneca Co., OH St. Charles Co., MO	1.05 (±.22) 1.56 (±1.81)	40 69	1.77	NS	Headley 1949
Solenopsis invicta	Madison & Monticello, GA	1.00 (± .00)	88			Fletcher et al. 1980
Formica pallide-fulva schau- fussi incerta	Hurley, MS Cheboygan, MI	19.44 (±21.5) 5.22 (±4.61)	25 23	8.14 4.49	<.001 <.001	Talbot, 1948
F. p. nitidiven- tris[d]		1.00 (± .00)	24			

[a]Only those cases where there are data for at least 10 complete colonies are reported. Data for colonies without a queen are not used.

[b]See also Elmes and Wardlaw 1982.

[c]These statistics not reported.

[d]Identifications as reported by Talbot (1948); Smith (1979) regards both as a single species.

Atta texana (Moser 1963; Echols 1966; Moser and Lewis 1981; Mintzer and Vinson 1985). The unicolonial/multicolonial dichotomy may obscure selective processes common in both groups. A unicolonial population structure represents an extreme on a continuum of possible population viscosities (sensu Hamilton 1964). The continual re-adoption of reproductives in unicolonial species, if coupled with a high degree of worker mixing, poses a potential problem for theories of eusociality based on relatedness. Under these conditions the effective degree of relatedness between a worker and aided queens vanishes, and the necessary inter-group competition component of kin selection (Wade 1980, 1982) does not exist, as workers do not aid selected 'family' members. Although continual re-adoption initially seems to enhance relatedness, dispersal of reproductives from family groups, permitting inter-familial competition, is mandatory for the evolution of altruism through kin selection (Pollock 1983). If workers preferentially aid their (half or full) sisters this problem is obviated. Mated queens of the secondarily polygynous acacia ant Pseudomyrmex venefica (Janzen 1973) are unable to enter any thorn of their parental tree, being most likely to gain access to the thorn from which they emerged. That workers repulse other queens suggests that the former remain faithful to specific thorns, for otherwise a worker would not know which queen 'belonged' in a thorn. Halliday (1983:53) describes a similar internal structure for unicolonial populations of I. purpureus where "each of [the] queens acts as a centre from which workers diffuse outward making smaller and smaller contributions to the populations of nests at increasing distances from their home nest." Unicolonial colonies may thus exhibit a more discrete internal structure than heretofore assumed, with patterns of worker migration revealing a population structure not dissimilar to that of multicolonial species, where discrete nest territories exist.

CONCLUSIONS

Multiple, functional queens in ant nests,
either during colony foundation or later in colony
development, is an important variation on either
the "wasp model," where queen subordination is
common, or the standard "haplometrosis model" of
ant colonies. While haplometrosis and monogyny
are common in ants, they are not ubiquitous;
pleometrosis and polygyny are common, even
abundant, in many species (Table 7.1, 7.2). Early
arguments concerning the role of the environment
in determining patterns of colony dynamics (for
example, Hölldobler and Wilson 1977) have been
followed by increasingly greater refinements
concerning, especially, the role of limiting
resources (viz. nest sites) in selecting for
pleometrosis and/or polygyny. Territoriality of
adult nests, when coupled with a clumped
distribution of starting nests, may represent a
common selection regime for pleometrosis. Little
is known, however, of the comparative population
structure for both adult and starting colonies in
species displaying alternative methods of colony
foundation. Related questions, such as alate
dispersal distance, degree of panmixia, and
colony/population-wide sex ratios also deserve
more attention. When coupled with abiotic factors
affecting nestsite quality, such characteristics
of the mating system may permit prediction as to
when starting colonies will be clumped and hence
subject to significant intra-species competition.
Divergent selective regimes for pleometrosis
may yield differing forms of queen interaction.
When selection for pleometrosis precludes
oligogyny, queens may eliminate each other once
the advantages of queen mutualism are at an end.
Oligogyny itself, besides limiting queen conflict,
should prevent queen parasitism when workers
exhibit fidelity to a single queen, either through
direct recognition or site specificity. Indeed,
the distinction between unicolonial and
multicolonial species may be of limited value when
workers do not migrate throughout their colony but
rather remain faithful to their site of
eclosion. Such site specificity is operationally
similar to multicolonial territoriality; when

queens are oligogynous, reduced migration raises the degree of relatedness between a queen and her workers. The effective operation of kin selection may require just such an intra-colonial population structure under 'unicolonial' secondary polygyny.

SUMMARY

While most ant colonies are founded by a single queen, "haplometrosis," colony foundation by several queens, "pleometrosis," occurs in a number of species. Cofoundresses regularly associate with non-relatives at colony foundation in several species, providing a contrast to pleometrotic wasps and an opportunity to examine cofoundress interaction in the absence of kin selection. Here we examine some of the selective forces yielding pleometrosis and their impact on cofoundress aggression/tolerance. We then review some of the literature on polygyny in adult ant colonies and argue that polygyny is more common than generally recognized. When polygyny is the product of pleometrosis, intracolony queen distribution may be a function of the original selective forces behind pleometrosis. Among unicolonial species, kin selection may itself require spacing between queens coupled with worker site fidelity; the population structure of unicolonial 'colonies' may not be that dissimilar from that of multicolonial species.

ACKNOWLEDGMENTS

Portions of the work reported here were supported by National Science Foundation grant DEB-8207052 to SWR.

REFERENCES

Adams, C.T., Banks, W.A., and Plumley, J.K.
1976. Polygyny in the tropical fire ant,
Solenopsis geminata with notes on the
imported fire ant, Solenopsis invicta. The
Florida Entomologist 59:411-415.
Alloway, T.M. 1980. The origins of slavery in
leptothoracine ants (Hymenoptera:
Formicidae). American Naturalist 115:247-
261.
Alloway, T.M., Buschinger, A., Talbot, M., Stuart,
R., and Thomas, C. 1982. Polygyny and
polydomy in three North American species of
the genus Leptothorax Mayr (Hymenoptera:
Formicidae). Psyche 89:249-272.
Baroni-Urbani, C. 1968a. Monogyny in ant
societies. Zoologischer Anzeiger 181:269-
277.
Baroni-Urbani, C. 1968b. Domination et monogynie
fonctionnelle dans une société digynique de
Myrmecina graminicola Latr. Insectes Sociaux
15:407-412.
Baroni-Urbani, C., and Soulié, J. 1962.
Monogynie chez la fourmi Crematogaster
scutellaris (Hymenoptera: Formicoidea).
Bulletin de la Société d'Histoire Naturelle
de Toulouse 97:29-34.
Bartz, S.H., and Hölldobler, B. 1982. Colony
founding in Myrmecocystus mimicus Wheeler
(Hymenoptera: Formicidae) and the evolution
of foundress associations. Behavioral
Ecology and Sociobiology 10:137-147.
Berkelhamer, R.C. 1980. Reproductive strategies
in ants: comparison of single-queened versus
multiple-queened species in the subfamily
Dolichoderinae. Ph.D. diss., University of
California, Berkeley.
Berkelhamer, R.C. 1984. An eletrophoretic
analysis of queen number in three species of
dolichoderine ants. Insectes Sociaux 31:132-
141.
Bernard, F. 1958. Résultants de la concurrence
naturelle chez les fourmis terricoles de
France et d'Afrique du Nord: évaluation
numérique des societes dominantes. Bulletin
de la Société d'Histoire Naturelle de

210

l'Afrique du Nord 49:302-356.
Beulig, M.L., and Janzen, D.H. 1969. Variation in behavior among obligate acacia-ants from the same colony (Pseudomyrmex nigrocincta). Journal of the Kansas Entomological Society 42:58-67.
Brian, M.V. 1965. Social insect populations. London: Academic Press.
Brian, M.V. 1983. Social insects. London: Chapman and Hall.
Buschinger, A. 1966. Untersuchungen an Harpagoxenus sublaevis Nyl. (Hymenoptera, Formicidae). I. Freilandbeobachtungen zu Verbreitung und Lebensweise. Insectes Sociaux 13:5-16.
Buschinger, A. 1968. Mono- und Polygynie bei Arten der Gattung Leptothorax Mayr (Hymenoptera: Formicidae). Insectes Sociaux 15:217-226.
Buschinger, A. 1974. Monogynie und Polygynie in Insektensozietaten. In Sozialpolymorphismus bei Insekten, ed. G.H. Schmidt, 862-896. Stuttgart: Wissenschaftliche Verlagsgesellschaft.
Buschinger, A. 1979a. Doronomyrmex pocahontas n. sp., a parasitic ant from Alberta, Canada (Hym. Formicidae). Insectes Sociaux 26:216-222.
Buschinger, A. 1979b. Functional monogyny in the American guest ant Formicoxenus hirticornis (Emery) (= Leptothorax hirticornis), (Hym, Form.). Insectes Sociaux 26:61-68.
Buschinger, A., and Alloway, T.M. 1977. Population structure and polymorphism in the slave-making ant Harpagoxenus americanus (Emery) (Hymenoptera: Formicidae). Psyche 84:233-242.
Buschinger, A., and Alloway, T.M. 1978. Caste polymorphism in Harpagoxenus canadensis M.R. Smith (Hym., Formicidae). Insectes Sociaux 25:339-350.
Buschinger, A., Francoeur, A., and Fischer, K. 1980. Functional monogyny, sexual behavior, and karyotype of the guest ant, Leptothorax provancheri Emery (Hymenoptera, Formicidae). Psyche 87:1-12.
Buschinger, A., and Winter, U. 1976.

Funktionelle Monogynie bei der Gastameise
Formicoxenus nitidulus (Nyl.) (Hym.,
Form.). Insectes Sociaux 23:549-558.
Buschinger, A., and Winter, U. 1983. Population
studies of the dulotic ant, Epimyrma ravouxi,
and the degenerate slavemaker, E. krussei
(Hymenoptera: Formicidae). Entomologia
Generalis 8:251-266.
Byron, P.A., Byron, E.R., and Bernstein, R.A.
1980. Evidence of competition between two
species of desert ants. Insectes Sociaux
27:351-360.
Clark, J. 1951. The Formicidae of Australia.
Vol. 1 Subfamily Myrmeciinae. Melbourne:
Morris & Walker Pty. Ltd.
Cory, E.N., and Haviland, E. 1938. Population
studies of Formica exsectoides. Annals of
the Entomological Society of America 31:50-
56.
Craig, R., and Crozier, R.H. 1979. Relatedness
in the polygynous ant Myrmecia pilosula.
Evolution 33:335-341.
Creighton, W.S. 1950. The ants of North
America. Bulletin of the Museum of
Comparative Zoology, Harvard University
104:1-585.
Creighton, W.S., and Gregg, R.E. 1954. Studies
on the habits and distribution of Cryptocerus
texanus Santschi (Hymenoptera:
Formicidae). Psyche 61:41-57.
Crozier, R.H. 1979. Genetics of sociality. In
Social insects, ed. H.R. Hermann, Vol. 1,
223-286. New York: Academic Press.
Echols, H.W. 1966. Compatibility of separate
nests of Texas leaf-cutting ants. Journal of
Economic Entomology 59:1299-1300.
Elmes, G.W. 1973. Observations on the density of
queens in natural colonies of Myrmica rubra
L. (Hymenoptera: Formicidae). Journal of
Animal Ecology 42:761-771.
Elmes, G.W. 1974a. The effect of colony
population size in three species of Myrmica
(Hymenoptera: Formicidae). Insectes Sociaux
21:213-230.
Elmes, G.W. 1974b. Colony populations of Myrmica
sulcinodis Nyl. (Hym. Formicidae). Oecologia
(Berlin) 15:337-343.

212

Elmes, G.W. 1980. Queen numbers in colonies of ants of the genus Myrmica. Insectes Sociaux 27:43-60.

Elmes, G.W., and Wardlaw, J.C. 1982. A population study of the ants Myrmica sabuleti and Myrmica scabrinodis, living at two sites in the South of England. I. A comparison of colony populations. Journal of Animal Ecology 51:651-664.

Evesham, E.J.M. 1984. The attractiveness of workers towards individual queens of the polygynous ant Myrmica rubra L. Biology of Behaviour 9:144-156.

Evesham, E.J.M. 1985. A video-based study of the role of workers in the movement of queens in the polygynous ant Myrmica rubra. Entomologia Experimentalis et Applicata 37:143-148.

Finnegan, R.J. 1977. Establishment of a predacious red wood ant, Formica obscuripes (Hymenoptera: Formicidae), from Manitoba to eastern Canada. Canadian Entomologist 109:1145-1148.

Fletcher, D.J.C. 1983. Three newly discovered polygynous populations of the fire ant, Solenopsis invicta, and their significance. Journal of the Georgia Entomological Society 18:538-543.

Fletcher, D.J.C., Blum, M.S., Whitt, T.V., and Temple, N. 1980. Monogyny and polygyny in the fire ant, Solenopsis invicta. Annals of the Entomological Society of America 73:658-661.

Fowler, H.G., and Roberts, R.B. 1983. Anomalous social dominance among queens of Camponotus ferrugineus (Hymenoptera: Formicidae). Journal of Natural History 17:185-187.

Gamboa, G.J. 1974. Surface behavior of the leaf-cutter ant Acromyrmex versicolor versicolor Pergande (Hymenoptera: Formicidae). Masters thesis, Arizona State University, Tempe, Arizona.

Glancey, B.M., Craig, C.H., Stringer, C.E., and Bishop, P.M. 1973. Multiple fertile queens in colonies of the imported fire ant, Solenopsis invicta. Journal of the Georgia Entomological Society 8:237-238.

Glancey, B.M., Stringer, C.E., Craig, C.H., and
     Bishop, P.M. 1975. An extraordinary case of
     polygyny in the red imported ant. Annals of
     the Entomological Society of America 68:922.
Gray, B. 1971. Notes on the biology of the ant
     species Myrmecia dispar (Clark)
     (Hymenoptera: Formicidae). Insectes Sociaux
     18:71-80.
Gray, B. 1974. Nest structure and populations of
     Myrmecia (Hymenoptera: Formicidae) with
     observations on the capture of prey.
     Insectes Sociaux 21:107-120.
Greenberg, L., Fletcher, D.J.C., and Vinson,
     S.B. 1985. Differences in worker size and
     mound distribution in monogynous and
     polygynous colonies of the fire ant
     Solenopsis invicta Buren. Journal of the
     Kansas Entomological Society 58:9-18.
Greenslade, P.J.M. 1971. Phenology of three ant
     species in the Solomon Islands. Journal of
     the Australian Entomological Society 10:241-
     252.
Hage, N., and Hanna, N.H.C. 1975. Contribution a
     l'etude de la biologie et de la polygynie de
     la fourmi Tapinoma simrothi phoenicium
     Emery. Comptes Rendu de l'Academie des
     Sciences, Serie D, Paris 281:1003-1005.
Halliday, R.B. 1983. Social organization of the
     meat ant Iridomyrmex purpureus analyzed by
     gel electrophoresis of enzymes. Insectes
     Sociaux 30:45-56.
Hamilton, W.D. 1964. The genetical evolution of
     social behavior, I and II. Journal of
     Theoretical Biology 7:1-52.
Haskins, C.P., and Haskins, E.F. 1950. Notes on
     the biology and social behavior of the
     archaic ponerine ants of the genera Myrmecia
     and Promyrmecia. Annals of the Entomological
     Society of America 43:461-491.
Headley, A.E. 1943. Population studies of two
     species of ants, Leptothorax longispinosus
     Roger and Leptothorax curvispinosus Mayr.
     Annals of the Entomological Society of
     America 36:743-753.
Headley, A.E. 1949. A population study of the
     ant Aphaenogaster fulva ssp. aquia Buckley.
     Annals of the Entomological Society of

214

America 42:265-272.
Herbers, J.M. 1984. Queen-worker conflict and eusocial evolution in a polygynous ant species. _Evolution_ 38:631-643.
Herbers, J.M. 1986. Effects of ecological parameters on queen number in _Leptothorax longispinosus_ (Hymenoptera: Formicidae). _Journal of the Kansas Entomological Society_ 59:675-686.
Herbers, J.M. 1987. Nest site limitation and facultative polygyny in the ant _Leptothorax longispinosus_. _Behavioral Ecology and Sociobiology_. 19:115-122.
Higashi, S. 1979. Polygyny, nest budding and internest mixture of individuals in _Formica (Serviformica) japonica_ Motschulsky at Ishikari shore. _Kontyû_ 47:381-389.
Higashi, S. 1983. Polygyny and nuptial flight of _Formica (Formica) yessensis_ Forel at Ishikari Coast, Hokkaido, Japan. _Insectes Sociaux_ 30:287-297.
Hölldobler, B. 1962. Zur Frage der Oligogynie bei _Camponotus ligniperda_ Latr. und _Camponotus herculeanus_ L. (Hym. Formicidae). _Zeitschrift für Angewandte Entomologie_ 49:337-352.
Hölldobler, B. 1976a. Recruitment behavior, home range orientation and territoriality in harvester ants, _Pogonomyrmex_. _Behavioral Ecology and Sociobiology_ 1:3-44.
Hölldobler, B. 1976b. Tournaments and slavery in a desert ant. _Science_ 192:912-914.
Hölldobler. B. 1981. Foraging and spatiotemporal territories in the honey ant _Myrmecocystus mimicus_ Wheeler (Hymenoptera: Formicidae). _Behavioral Ecology and Sociobiology_ 9:301-314.
Hölldobler, B., and Carlin, N.F. 1985. Colony founding, queen dominance and oligogyny in the Australian meat ant _Iridomyrmex purpureus_. _Behavioral Ecology and Sociobiology_ 18:45-58.
Hölldobler, B., and Wilson, E.O. 1977. The number of queens: an important trait in ant evolution. _Naturwissenschaften_ 64:8-15.
Ito, M. 1973. Seasonal population trends and nest structure in a polydomous ant, _Formica_

(Formica) yessensis Forel. Journal of the
Faculty of Science Hokkaido University,
Series VI (Zoology) 19:270-292.
Janzen, D.H. 1973. Evolution of polygynous
obligate acacia-ants in western Mexico.
Journal of Animal Ecology 42:727-750.
Jayasuriya, A.K., and Traniello, J.F.A. 1986.
The biology of the primitive ant Aneuretus
simoni (Emery) (Formicidae: Aneuretinae).
I. Distribution, abundance, colony structure
and foraging ecology. Insectes Sociaux
32:363-374.
Jeanne, R.L. 1972. Social biology of the
Neotropical wasp Mischocyttarus drewseni.
Bulletin of the Museum of Comparative
Zoology, Harvard University, Cambridge,
Massachusetts 144:63-150.
Kannowski, P.B. 1967. Colony populations of two
species of Dolichoderus (Hymenoptera:
Formicidae). Annals of the Entomological
Society of America 60:1246-1252.
Kannowski, P.B. 1970. Colony populations of five
species of Myrmica (Hymenoptera:
Formicidae). Proceedings of the
Entomological Society of America, North
Central Branch 25:119-125.
Kloft, W.J., Wilkenson, R.C., Whitcomb, W.H., and
Kloft, E.S. 1973. Formica integra
(Hymenoptera: Formicidae) I. Habitat, nest
construction, polygyny, and biometry.
Florida Entomologist 52:67-76.
Lavigne, R.J. 1969. Bionomics and nest structure
of Pogonomyrmex occidentalis (Hymenoptera:
Formicidae). Annals of the Entomological
Society of America 62:1166-1175.
Letendre, M., and Pilon, J.G. 1972. Écologie des
populations de Leptothorax longispinosus
Roger et Stenamma diecki Emery dans les
peuplements forestiers des basses
Laurentides, Québec (Hymenoptera:
Formicidae). Le Naturaliste Canadien 99:73-
82.
Litte, M. 1979. Mischocyttarus flavitarsis in
Arizona: social and nesting biology of a
polistine wasp. Zeitschrift für
Tierpsychologie 50:282-312.
MacKay, W.P. 1981. A comparison of the nest

216

phenologies of three species of Pogonomyrmex harvester ants (Hymenoptera: Formicidae). Psyche 88:25-47.

Marikovsky, P.I. 1963. The ants Formica sanguinea Latr. as pillagers of Formica rufa nests. Insectes Sociaux 10:119-128.

Markin, G.P. 1970. The seasonal life cycle of the Argentine ant, Iridomyrmex humilis (Hymenoptera: Formicidae), in southern California. Annals of the Entomological Society of America 63:123-124.

Markin, G.P., Collins, H.L., and Dillier, J.H. 1972. Colony founding by queens of the red imported fire ant, Solenopsis invicta. Annals of the Entomological Society of America 65:1053-1058.

Markin, G.P., Dillier, J.H., and Collins, H.L. 1973. Growth and development of colonies of the red imported fire ant, Solenopsis invicta. Annals of the Entomological Society of America 66:803-808.

Martin, M.M., Carls, G.A., Hutchins, R.F.N., MacConnell, J.G., Martin, J.S., and Steiner, O.D. 1967. Observations on Atta columbica tonsipes (Hymenoptera: Formicidae). Annals of the Entomological Society of America 60:1329-1330.

Metcalf, R.A., and Whitt, G.S. 1977. Relative inclusive fitness in the social wasp Polistes metricus. Behavioral Ecology and Sociobiology 2:353-360.

Mintzer, A. 1979. Colony founding and pleometrosis in Camponotus (Hymenoptera: Formicidae). Pan-Pacific Entomologist 55:81-89.

Mintzer, A., and Vinson, S.B. 1985. Cooperative colony foundation by females of the leafcutting ant Atta texana in the laboratory. Journal of the New York Entomological Society 93:1047-1051.

Mizutani, A. 1981. On the two forms of the ant Myrmica ruginodis Nylander (Hymenoptera: Formicidae) from Sapporo and its vicinity, Japan. Japanese Journal of Ecology 31:131-137.

Mizutani, A., and Imamura, S. 1980. Population and nest structure in the ant Aphaenogaster

japonica Forel, in Sapporo, Japan. Kontyû
48:241-247.
Moser, J.C. 1963. Contents and structure of _Atta_
texana nest in summer. Annals of the
Entomological Society of America 56:286-291.
Moser, J.C. 1967. Mating activities of _Atta_
texana (Hymenoptera: Formicidae). Insectes
Sociaux 16:295-312.
Moser, J.C., and Lewis, J.R. 1981. Multiple nest
queens of _Atta texana_. Turrialba 31:256-257.
Nickerson, J.C., Cromroy, H.L., Whitcomb, W.H.,
and Cornell, J.A. 1975. Colony organization
and queen numbers in two species of
Conomyrma. Annals of the Entomological
Society of America 68:1083-1085.
Noonan, K.M. 1981. Individual strategies of
inclusive-fitness-maximizing in _Polistes_
fuscatus foundresses. In Natural selection
and social behavior, ed. R.D. Alexander and
D.W Tinkle, 18-44. New York: Chiron Press.
Onoyama, K. 1981. Brood rearing by colony
founding queens of the harvester ant, _Messor_
aciculatus. Kontyû 49:624-640.
Oster, G.F., and Wilson, E.O. 1978. Caste and
ecology in the social insects. Princeton,
New Jersey: Princeton University Press.
Pamilo, P. 1981. Genetic organization of _Formica_
sanguinea populations. Behavioral Ecology
and Sociobiology 9:45-50.
Pamilo, P. 1982. Genetic population structure in
polygynous _Formica_ ants. Heredity 48:95-106.
Pamilo, P. 1983. Genetic differentiation within
subdivided populations of _Formica_ ants.
Evolution 37:1010-1022.
Pamilo, P., and Rosengren, R. 1984. Evolution of
nesting strategies of ants: genetic evidence
from different population types of _Formica_
ants. Biological Journal of the Linnean
Society 21:331-348.
Pamilo, P., and Varvio-Aho, S.-L. 1979. Genetic
structure of nests in the ant _Formica_
sanguinea. Behavioral Ecology and
Sociobiology 6:91-98.
Pardi, L. 1948. Dominance order in _Polistes_
wasps. Physiological Zoology 21:1-13.
Pearson, B. 1982. Relatedness of normal queens
(macrogynes) in nests of the polygynous ant

Myrmica rubra L. *Evolution* 36:107-112.

Pfennig, D.D., Gamboa, G.J., Reeve, H.K., Reeve, J.S., and Ferguson, I.D. 1983. The mechanism of nestmate discrimination in social wasps (*Polistes*, Hymenoptera: Vespidae). *Behavioral Ecology and Sociobiology* 13:299-305.

Pollock, G.B. 1983. Population viscosity and kin selection. *American Naturalist* 122:817-829.

Pollock, G.B., and Rissing, S.W. 1985. Mating season and colony foundation of the seed-harvester ant, *Vermessor pergandei*. *Psyche* 92:125-134.

Pontin, A.J. 1978. The numbers and distribution of subterranean aphids and their exploitation by the ant *Lasius flavus* (Fabr.). *Ecological Entomology* 3:203-207

Pricer, J.L. 1908. The life history of the carpenter ant. *Biological Bulletin* 14:177-218.

Provost, É. 1979. Étude de la fermeture de la société de fourmis chez diverse espèces de *Leptothorax* et chez *Camponotus lateralis* (Hyménoptères: Formicidae). *Comptes Rendu de l'Académie des Sciences*, Série D, Paris 288:429-432.

Rettenmeyer, C.W., and Watkins, II, J.F. 1978. Polygyny and monogyny in army ants (Hymenoptera: Formicidae). *Journal of the Kansas Entomological Society* 51:581-591.

Rissing, S.W., and Pollock, G.B. 1986. Social interaction among pleometrotic queens of *Veromessor pergandei* (Hymenoptera: Formicidae) during colony foundation. *Animal Behaviour* 34:226-233.

Rissing, S.W., and Pollock, G.B. 1987. Queen aggression, pleometrotic advantage and brood raiding in the ant *Veromessor pergandei* (Hymenoptera: Formicidae). *Animal Behaviour* 35:975-981.

Rissing, S.W., Johnson, R.A., and Pollock, G.B. 1986. Natal nest distribution and pleometrosis in the desert leaf-cutter ant *Acromyrmex versicolor* (Pergande) (Hymenoptera: Formicidae). *Psyche* 93:177-186.

Ross, K.G., and Fletcher, D.J.C. 1985.

Comparative study of genetic and social structure in two forms of the fire ant Solenopsis invicta (Hymenoptera: Formicidae). Behavioral Ecology and Sociobiology 17:349-356.

Ryti, R.T., and Case, T.J. 1984. Spatial patterns and diet overlap between colonies of three desert ant species. Oecologia (Berlin) 62:401-404.

Skaife, S.H. 1961 The black sugar ant Camponotus werthi Forel (Hymenoptera: Formicidae). Journal of the Entomological Society of South Africa 24:110-124.

Smith, D.R. 1979. Superfamily Formicoidea. In Catalog of Hymenoptera in America north of Mexico. ed. K.V. Krombein (and others), vol. 2, Washington, D.C.: Smithsonian Institution Press.

Smith, M.R. 1928. The biology of Tapinoma sessile Say, an important house-infesting ant. Annals of the Entomological Society of America 21:307-330.

Soulié, J. 1964. Le contrôle par les ouvrières da la monogynie des colonies chez Sphaerocrema striatula (Myrmicidae, Cremastogastrini). Insectes Sociaux 11:383-388.

Stumper, R. 1962. Sur un effet de groupe chez les femelles de Camponotus vagus (Scopoli). Insectes Sociaux 9:329-333.

Taki, A. 1976. Colony founding of Messor aciculatum (Fr. Smith) (Hymenoptera: Formicidae) by single and grouped queens. Physiology and Ecology Japan 17:503-512.

Talbot, M. 1943. Population studies of the ant Prenolepsis imparis Say. Ecology 24:31-44.

Talbot, M. 1945. Population studies of the ant Myrmica schencki ssp. emeryana Forel. Annals of the Entomological Society of America 38:365-372.

Talbot, M. 1948. A comparison of two ants of the genus Formica. Ecology 29:316-325.

Talbot, M. 1951. Populations and hibernating conditions of the ant Aphaenogaster (Attomyrma) rudis Emery. Annals of the Entomological Society of America 44:302-307.

Talbot, M. 1954. Populations of the ant

Aphaenogaster (Attomyrma) treatae (Forel) on abandoned fields on the Edwin S. George Reserve. Contributions from the Laboratory of Vertebrate Biology of the University of Michigan 69:1-9.

Talbot, M. 1957. Population studies of the slave-making ant Leptothorax duloticus and its slave, Leptothorax curvispinosus. Ecology 38:449-456.

Talbot, M. 1965. Populations of ants in a low field. Insectes Sociaux 12:19-48.

Talbot, M. 1975. Habitats and populations of the ant Stenamma diecki Emery in Southern Michigan. Great Lakes Entomologist 8:241-244.

Terron, G. 1977. Évolution des colonies de Tetraponera anthracina Santschi (Formicidae Pseudomyrmecinae) avec reines. Bulletin Biologique de la France et de la Belgique 111:115-182.

Tevis, L. 1958. Interrelationships between the harvester ant Veromessor pergandei (Mayr) and some desert ephemerals. Ecology 39:695-704.

Torossian, C. 1960. La Biologie de la fourmi Dolichoderus quadripunctatus (Hyménoptère-Formicidae-Dolichoderinae). Insectes Sociaux 7:383-393.

Traniello, J.F.A. 1982. Population structure and social organization in the primitive ant Amblyopone pallipes (Hymenoptera: Formicidae). Psyche 89:65-80.

Tschinkel, W.R., and Howard, D.F. 1978. Queen replacement in orphaned colonies of the fire ant, Solenopsis invicta. Behavioral Ecology and Sociobiology 3:297-310.

Tschinkel, W.R., and Howard, D.F. 1983. Colony founding by pleometrosis in the fire ant, Solenopsis invicta. Behavioral Ecology and Sociobiology 12:103-113.

Wade, M.J. 1980. Kin selection: its components. Science 210:665-667.

Wade, M.J. 1982. The effect of multiple inseminations on the evolution of social behaviors in diploid and haplo-diploid organisms. Journal of Theoretical Biology 95:351-368.

Waloff, N. 1957. The effect of the number of

queens of the ant Lasius flavus (Fab.) (Hym.,
Formicidae) on their survival and on the rate
of development of the first brood. Insectes
Sociaux 4:391-408.
Ward, P.S. 1981. Ecology and life history of the
Rhytidoponera impressa group (Hymenoptera:
Formicidae). II. Colony origin, seasonal
cycles, and reproduction. Psyche 88:109-126.
Weber, N.A. 1972. Gardening ants, the attines.
Memoirs of the American Philosophical Society
92:1-146.
Went, F.W., Wheeler, G.C., and Wheeler, J.
1972. Feeding and digestion in some ants
(Veromessor and Manica). Bioscience 22:82-8.
Wesson, L.G. 1939. Contribution to the natural
history of Harpagoxenus americanus Emery.
Transactions of the American Entomological
Society 65:97-122.
West, M.J. 1967. Foundress associations in
polistine wasps: dominance hierarchies and
the evolution of social behavior. Science
157:1584-1585.
West Eberhard, M.J. 1969. The social biology of
polistine wasps. Miscellaneous Publications
of the Museum of Zoology, University Michigan
140:1-101.
Wheeler, D.E. 1982. Soldier determination in the
ant, Pheidole bicarinata. Ph.D. diss., Duke
University.
Wheeler, G.C., and Wheeler, J. 1963. The ants of
North Dakota. Grand Forks: University of
North Dakota Press.
Wheeler, J., and Rissing, S.W. 1975a. Natural
history of Veromessor pergandei. II.
Behavior. Pan-Pacific Entomologist 51:303-
314.
Wheeler, J., and Rissing, S.W. 1975b. Natural
history of Veromessor pergandei. I. The
nest. Pan-Pacific Entomologist 51:205-216.
Wheeler, W.M. 1910. Ants: their structure,
development and behavior. New York:
Columbia University Press.
Wheeler, W.M. 1917. The pleometrosis of
Myrmecocystus. Psyche 104:180-182.
Wheeler, W.M. 1933. Colony-founding among ants,
with an account of some primitive Australian
species. Cambridge, Massachusetts: Harvard

University Press.

Wilson, E.O. 1963. Social modifications related to rareness in ant species. Evolution 17:249-253.

Wilson, E.O. 1974. The population consequences of polygyny in the ant Leptothorax curvispinosus. Annals of the Entomological Society of America 67:781-786.

Wilson, N.L., Dillier, J.H., and Markin, G.P. 1971. Foraging territories of imported fire ants. Annals of the Entomological Society of America 64:660-665.

Yamauchi, K., Kinomura, K., and Miyake, S. 1981. Sociobiological studies of the polygynic ant Lasius sakagamii. I. General features of its polydomous system. Insectes Sociaux 28:279 296.

Yamauchi, K., Kinomura, K., and Miyake, S. 1982. Sociobiological studies of the polygynic ant Lasius sakagamii. II. Production of colony members. Insectes Sociaux 29:164-174.

8

Behavioral and Biochemical Variation in the Fire Ant, *Solenopsis invicta*

Robert K. Vander Meer

The fire ant, Solenopsis invicta Buren, was
accidentally introduced into the southern United
States in the 1930s from South America. It is
most probable that the United States S. invicta
population is derived from a limited introduced
gene pool, which may dampen within-colony and
colony to colony variation (Tschinkel and
Nierenberg 1983). Whether this is indeed the case
has yet to be determined; however, the fire ant
genetic story continues to become more
complicated. Prior to S. invicta's introduction a
black form had been introduced (early 1900s) from
South America. The two forms were originally
classified as racial forms of the same species
(Wilson 1953); however, each was subsequently
given species status (Buren 1972; black form =
Solenopsis richteri Forel, red form = S. invicta)
based on morphological differences, lack of
hybridization in the United States, and phenetic
invariability in the two forms. Recently
extensive hybridization between the two forms has
been detected biochemically (Vander Meer et al.
1985) and confirmed using isozymes (Ross et al.
1987). The hybrid has the color and morphology of
S. richteri. As a consequence we cannot be
certain of the integrity of the United States S.
invicta population. In addition Ross et al.
(1985) found a substantial amount of allozyme
variability in the United States S. invicta
population, which indicates that there may have
been multiple introductions. In any event, we do
not know the precise history of the introduced

223

Solenopsis species and we know even less about the
ants in their native South America. Based on
personal experience and the work of others it is
clear that *S. invicta* has maintained enough
variability to add gray hairs to many
scientists. The following discussion outlines
where variability has been identified.

Female categories within a colony are
illustrated in Figure 8.1. Single queen *S.
invicta* colonies are founded independently by
either single or multiple newly mated queens,
which lead to primary and secondary monogyny,
respectively (Hölldobler and Wilson 1977a;
Tschinkel and Howard 1983). It should be noted
that increasing numbers of polygynous *S. invicta*
colonies have been discovered in recent years
(Glancey et al. 1975; Fletcher 1983; Fletcher et
al. 1980; Lofgren and Williams 1984). Whether
this is a recent phenomenon or a result of
heightened awareness is unknown. This paper will
not consider polygynous *S. invicta* colonies, since
monogynous colonies are complicated enough. The
colony life cycle can be divided into three major
stages: (A) colony founding, (B) the ergonomic
stage, (C) the reproductive stage (Oster and
Wilson 1978). Each of these will be dealt with in
the following discussion.

THE COLONY FOUNDING STAGE

At colony founding the newly mated fire ant
queen constructs a nuptial chamber and rears the
first brood in seclusion and without foraging
(claustral colony founding). She relies on food
and energy reserves built up prior to her mating
flight. In addition, her no longer required wing
muscles are histolyzed to provide a needed source
of peptides and amino acids for egg production and
general maintenance (Toom et al. 1976a). The
first pupae that eclose are called nanitic or
minim workers, and are the smallest workers
produced by a colony (Dumpert 1981; Porter and
Tschinkel 1986). It has been assumed that in most
social insects the production of many nanitic
workers versus a few larger workers has adaptive
advantages for successful colony foundation (Oster

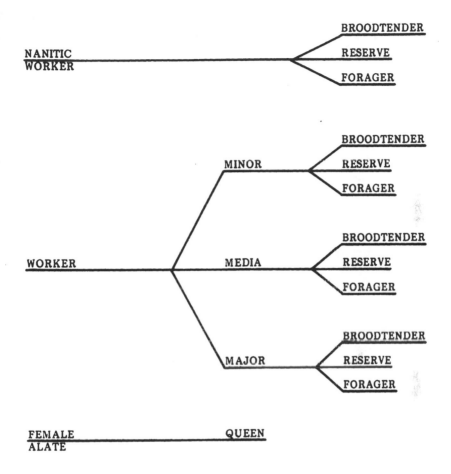

Figure 8.1.   Solenopsis invicta female castes and
     sub-castes: sources of variability.

and Wilson 1978).   This was demonstrated for S.
invicta by Porter and Tschinkel (1986), when they
showed that founding colonies with nanitic workers
produced significantly more brood than modified
colonies with equivalent biomasses of larger
workers.   The number of workers rather than their
size was the critical factor influencing brood
production (Porter and Tschinkel 1986).   Nanitic
workers are of uniform size (headwidths are 0.49 ±
0.00 mm [mean ± SD] N = 24) and unlike their
counterparts in a mature colony, who have the
luxury of undergoing age polyethism (Mirenda and
Vinson 1981), they must immediately tend brood,

maintain the nest, and forage for food (Oster and Wilson, 1978). It has also been noted (Oster and Wilson, 1978:46) that "one of the most general rules of behavior in all social insects that the members of incipient colonies are more timid than the members of large colonies." The distinct monomorphic size and difference in behavior of nanitic workers compared to mature colony workers suggests that they may be considered a separate caste. This hypothesis was further confirmed by chemical analysis of venom alkaloids (2-alkyl or alkenyl-6-methyl piperidines, Brand et al., 1972) that clearly demonstrated distinct venom alkaloid patterns between nanitic workers and mature colony workers (Vander Meer 1986a; Vander Meer, unpublished data). These differences were demonstrated to be unrelated to worker size. The small size of nanitic workers has been attributed to the poor diet provided by the queen (Dumpert 1981). Experiments designed to test this hypothesis led to the conclusion that nanitic workers are produced regardless of the diet they receive in the incipient colony (Vander Meer, unpublished data). The conclusion is that nanitic workers are a distinct but transient caste, only required during the colony founding stage. Vander Meer (unpublished) has biochemical evidence that nanitic characteristics carry on in the maturing colony for a much longer period of time than defined by the development of a queen's first brood. This has important implications regarding potential variability of colonies and workers for both biochemical and behavioral studies.

THE ERGONOMIC STAGE

After the first workers emerge the colony begins the ergonomic stage, which consists of the buildup of the colony work force. The size of a colony is limited by the egg production of the queen, which is in turn dependent on the rate of food flow supplied to her by the workers (aided perhaps by contributions of the fourth instar larvae). Since food flow is dependent on the size of the worker force, egg production and therefore colony size spiral up to the maximum permitted by

the reproductive capacity of the individual queen
and extant environmental conditions. The rapid
growth of a colony is facilitated by a division of
labor strategy (Oster and Wilson, 1978). In S.
invicta division of labor is based on a blend of
age polyethism and caste based on physical size.

Size Subcastes

After the colony founding stage, worker size
distribution changes, with larger workers
developing as the colony matures (Wood and
Tschinkel 1981). In mature colonies workers
exhibit slight polymorphism and occur in a
continuous range of sizes; however, minor, media,
and major size categories have been arbitrarily
assigned to workers based on head width
measurements (Wilson 1978). Although worker size
distributions are related to colony size, the
distribution is not homeostatically regulated
(Porter and Tschinkel 1985a). There is a trend
toward specialization in the larger size
classes. Mirenda and Vinson (1981) showed that
worker size was correlated with the size of
particle carried (for nest maintenance or food
retrieval). The smaller workers of the same age
were more involved in brood care than the larger
workers. Experiments with artificial monomorphic
S. invicta colonies also showed that colonies
composed of only large workers produced little
brood compared to standard polymorphic colonies,
whereas colonies composed only of small workers
produced brood at the same rate as standard
polymorphic colonies (Porter and Tschinkel
1985b). However, small worker monomorphic
colonies were 10% less energy efficient in brood
production than corresponding polymorphic colonies
(Porter and Tschinkel 1985b). Analysis of venom
alkaloids from a range of mature colony worker
sizes showed a relationship between the percentage
of certain alkaloids and the size of the worker
(Vander Meer 1986a; Vander Meer, unpublished
data). In addition, the same size categories were
analyzed for total alkaloid (Figure 8.2). The
data show that although major workers (headwidth =
ca. 1.10 mm) are considerably larger than minor

228

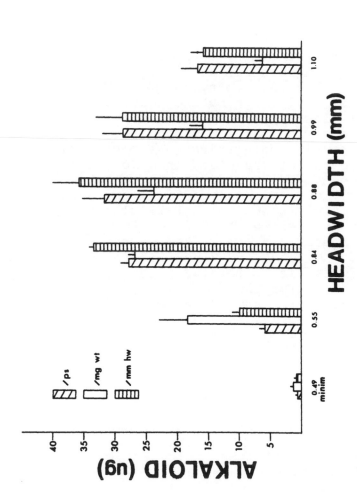

Figure 8.2. The total amount of alkaloid from extirpated glands of various S. invicta worker size categories. Analysis by gas chromatography using an internal standard and a Varian 401 data processor.

workers (headwidth = ca. 0.84 mm), they contain
less alkaloid. Based on the amount of alkaloid
available, major workers are less capable of
defending their nest than are minor workers. This
was demonstrated behaviorally by Wilson (1978),
who found that fire ant aggressive behavior toward
introduced foreign objects decreases with
increasing size and major workers do not function
as a soldier caste. Although S. invicta workers
are polymorphic with no distinct morphological
castes, worker size represents an important colony
variable when considering both biochemical and
behavioral studies.

Temporal Subcastes

Nanitic workers (the first adults produced by
a colony) must immediately carry out all colony
functions; however, mature workers exhibit age
related polyethism (Mirenda and Vinson 1981). Age
or temporal polyethism is a long established
characteristic of eusocial insects (Wilson
1971). Mirenda and Vinson (1981) identified three
fire ant subcastes: behaviorally distinct nurses
and foragers, and a less distinct intermediate
group called reserves. Callow workers spent most
of their time near the brood but gradually moved
to the periphery of the nest as they aged. Under
the laboratory conditions of the experiment,
workers that moved to the nest periphery were much
less active than those tending brood. These
inactive workers were called reserves, and
functioned in food transfer from forager to nurse
(Sorensen et al. 1985a). Under natural conditions
they may be active in nest maintenance and
construction. As reserves continued to age they
spent increasing amounts of time in the foraging
arena until near death, when they were
consistently found foraging (Mirenda and Vinson
1981). Although this is complicated enough, the
confounding size factor must also be considered.
Nurses were primarily mature minor workers with a
small proportion of media workers. Some workers,
particularly major workers, never tended brood and
very quickly moved to the nest periphery with the
reserves. Some of these major workers may serve

as repletes (Glancey et al. 1973). Mirenda and Vinson (1981) found that although there were significant average differences in task performance due to age, interindividual variance was the largest contributor to the total variance in time spent in certain locations and performing certain behavior. This type of individual variation can be due to differences in an individual's past experience, genetics, or other unknown factors. Although time is a convenient measure of the progression of a worker from nurse to forager, it is not accurate due to the large individual variation. Future efforts at devising methods for measuring an individual ant's physiological rather than temporal age could have a significant impact on our understanding of social insect behavior. This has been successfully done in the beetle, Oryctes rhinoceros L. (Vander Meer 1987), in which several behavioral acts (first flight and feeding) were strongly correlated with a beetle's weight at the time of the behavioral act and its adult emergent weight. In contrast, time was a poor predictor. Ants also have a fixed exoskeleton, which might make them amenable to a similar analysis.

The context in which a worker finds herself is another potentially confounding factor. Sorensen et al. (1985b) found that S. invicta nurses, reserves, and foragers exhibited a high degree of behavioral flexibility when posed with artificial elimination of all but their own subcaste in a colony.

One bright note of consistency was found amid the experimenter's nightmares discussed above. Quantitative and qualitative analyses of the venom alkaloids from the three defined temporal castes were not significantly different from each other when ants of approximately the same size were used in the analysis (Table 8.1). This means that, at least in terms of this particular biochemical character, fire ant workers are consistent over time. This has made venom alkaloids an important chemotaxonomic tool (Vander Meer 1986a; Vander Meer et al. 1985). Nurses have as much venom alkaloid capacity as foragers (Table 8.1), which are responsible for prey capture and colony defense because of the multiple functions of the

Table 8.1
Comparison of Solenopsis invicta temporal caste venom alkaloids[a,b]

Temporal caste	$C_{11:0}$	$C_{13:1}$	$C_{13:0}$	$C_{15:1}$	$C_{15:0}$	Total alkaloid (µg)
Brood tender	0.44 ±0.05	33.75 ±2.69	12.83 ±0.67	41.08 ±0.70	12.17 ±1.42	8.35 ±2.16
Reserve	0.33 ±0.01	31.42 ±1.61	13.38 ±0.49	42.44 ±0.87	12.52 ±0.86	6.30 ±1.76
Forager	0.88 ±0.30	34.96 ±1.15	12.29 ±0.27	41.39 ±0.77	10.84 ±0.37	7.80 ±1.23

[a]Mean percent total alkaloid ± SD; N = 5.
[b]No significant differences between the temporal castes for percent alkaloid or total alkaloid (Wilcoxon's 2-sample test for unpaired observation $\underline{P} > .05$).

poison sac contents and the sting apparatus. The
venom alkaloids have a wide variety of
physiological effects, such as hemolysis (Adrouny
et al. 1959), histamine release from mast cells
(Lind 1982), and antibiotic activity (Blum et al.
1958; Jouvenaz et al. 1972). Obin and Vander Meer
(1985) discovered a basic gaster vibration
behavior used by S. invicta nurses and foraging
workers. The effects of sting extension, venom
release, and gaster vibration are to disperse fine
droplets of venom into the surrounding
environment. For a forager this can have a
repellent effect on an antagonist, preventing
direct confrontation, as reported for other ant
species (Holldobler 1982; Adams and Traniello
1981). For a brood tending ant this mechanism
allows the dispersal of minute amounts of the
antibiotic alkaloids on the brood and within the
nest to help defend against omnipresent soil
bacteria and fungi (Obin and Vander Meer 1985).
Consequently, regardless of a worker's position
within a colony (nurse or forager), there is a
continuous requirement for the production and
dispersal of venom alkaloids.

Changes in behavioral sensitivity. Over many
years our laboratory has investigated and
developed S. invicta bioassays for queen-produced
pheromones (Vander Meer et al. 1980; Glancey et
al. 1982), brood tending pheromones (Glancey et
al. 1970), and trail pheromones (Jouvenaz et al.
1978; Vander Meer et al. 1981). The isolation of
pheromones requires a bioassay that offers a
quick, reliable, and definitive answer.
Consequently, for pheromone isolation we modified
the bioassays until the worker ant response was
optimized. Interestingly, brood tending workers
gave the most consistent results in many of the
bioassays. We have recently systematically
investigated the responsiveness of nurses,
reserves, and foragers to several pheromone
systems (Glancey et al., unpublished data).
Response to the queen attractant at five
concentrations was measured in an olfactometer
bioassay (Vander Meer et al. 1980). The results
show that the proportion of brood tenders
responding was significantly greater than
reserves, which responded significantly better

than foragers. These results can be readily rationalized, since brood tending workers are in closer association with the queen than are reserves or foragers, which could result in the sensitization of brood tenders to queen-produced pheromones. Based on our previous temporal caste definitions, reserves would be expected to fall somewhere between brood tenders and foragers, and they do. Foragers are the furthest away from the influence of the queen and should be the least sensitive to the queen-produced pheromones.

A second bioassay measured the time it took an individual worker placed on a pile of brood to pick up a piece of brood. The results showed significant differences between each of the three subcastes. Brood tenders responded more quickly than the other two subcastes. These results are not surprising, since again brood tending workers interact directly with the brood and might be expected to be sensitized to brood pheromones. As the workers age and move through the reserve and forager subcaste they would be expected to be less sensitive to brood pheromones because of being further away from brood interactions in both physical distance and time.

Another bioassay involved worker attraction to the recruitment part of the trail pheromone as measured by the response of worker ants to Dufour's gland volatiles in a Y-tube olfactometer (Vander Meer et al., in press). The results were similar to those obtained in the previous two bioassays. Brood tenders are significantly more sensitive to the trail recruitment pheromone than are reserves and foragers. Previous arguments were based on the assumption that association with a particular pheromone system promotes an increase in sensitivity to that pheromone. The expected order of responsiveness then would be forager greater than reserve, greater than brood tender. On the other hand, it could also be argued that continual contact with a particular pheromone system could lead to a decreased response due to habituation (Shorey 1976). This hypothesis would predict the experimental order of responsiveness for the trail pheromone, but the opposite results in the other two bioassays. The data for the three bioassays are consistent. Brood tenders are

the most sensitive temporal subcaste in all
bioassays, and the order of decreasing sensitivity
is in the direction of increasing age. The most
plausible explanation is also the least
complicated: as workers age they become less
sensitive to all pheromone systems as a
consequence of that catchall term, senescence.
The experimental results may be a reflection of
deterioration of the physiological and/or physical
systems (see Porter and Jorgensen 1981). The
conclusion from these experiments is that all
worker ants in a colony are not equal in their
behavioral responses to pheromones and perhaps
other stimuli. The responsiveness of workers to
pheromone systems in general diminishes with the
physiological age of the individual.

Variation in trail formation ability. The
fire ant is noted for its efficient foraging
strategy, which involves a combination of search
behavior, homing mechanisms, and the use of trail
pheromones for recruitment and trail orientation
(Wilson 1962). One of the most important elements
of this strategy is the homing and subsequent
trail formation to the food source. The majority
of competition for food resources comes from other
fire ant colonies and other ant species. We
measured the initial trail formation abilities of
several S. invicta laboratory colonies (all
laboratory colonies are derived from field
colonies or from newly mated queens collected from
the field). Colonies of approximately equal size
were transferred to clean trays. After one to
three hours a cockroach was placed 25 cm from the
colony cell and the time required for formation of
a discernable worker trail was recorded. This
bioassay measured a combination of (A) the
foraging worker's ability to find its way back to
the colony, (B) the foraging worker's ability to
deposit a chemical trail, and (C) the
responsiveness of reserve workers or other
foraging workers to the recruitment and
orientation pheromones (Vander Meer 1986b). Table
8.2 shows the remarkable variability we found in a
colony's trail formation ability. Within colony
variation was small compared with colony to colony
differences. In the laboratory the time it takes

to recruit workers to a food source is not important to the colony's survival, since there is no intraspecific or interspecific competition. However, in a field situation, SNAIL, the slowest colony to respond, would have a difficult time competing for resources with LIGHTNING, the quickest to respond, or any of the other colonies. The moral of these results is that at any given time colony behavioral responses are not equal. In addition to the general within colony differences in pheromone responsiveness associated with age polyethism, we have the further confounding colony to colony variation, which, in the case of initial trail formation, is considerable.

THE REPRODUCTIVE STAGE

Female alates are another female caste, morphologically and chemically different from

Table 8.2
Comparison of initial trail formation response times for nine *Solenopsis invicta* laboratory colonies

Colony name	Mean response (min)	SD	SE	Comp.[a]
Snail	21.28	2.98	1.49	A
Tortoise	13.40	4.71	2.35	B
Molasses	10.74	2.06	1.03	BC
Slowpoke	10.10	2.04	1.02	BCD
Mediocre	7.58	2.94	1.47	CDE
W.E. Coyote	6.83	1.71	0.85	DEF
Road-runner	4.78	1.76	0.88	EFG
Secretariat	2.88	0.53	0.27	G
Lightning	2.16	0.21	0.11	G

[a]Colonies with different letters are significantly different from each other ($P < .05$, $N = 4$, Newman-Keuls test).

workers (Brand et al. 1973). The size of alates,
measured by headwidth, is constant whether they
are from the laboratory or field (Table 8.3).
This indicates that the fixed morphology of the
adult female sexual caste is not influenced by
diet. However, the weight of alates increases as
they mature and store the energy and crop reserves
required for colony founding (Toom et al.
1976b). Invariably, spring-reared alates captured
when about to fly from their nest are of uniform
weight; however, overwintering alates captured
when about to leave on mating flights weighed one-
third less (Fletcher and Blum 1983a). A
comparison of mating and colony founding success
for these two groups has not been done, however it
would be predicted that the overwintering alates
would be less fit for colony founding. This would
definitely be true for haplometrotic colony
founding (founding by individuals) but maybe not
so critical in pleometrosis (founding in
groups). Perhaps there is a connection between
the development of polygynous S. invicta colonies
(Lofgren and Williams 1984) and the initial
interdependency of the founding queens. This
interdependency may occur in colonies started from
overwintered newly mated queens. An alate's
behavior depends on its physiological state, and
unless colonies in the field or laboratory are

Table 8.3
Solenopsis invicta female sexual head width
measurements

| Type | Source | Head width (mm) | | $\underline{N}$ |
		Mean	SD	
Alate	Laboratory	1.43	±0.00	3
Alate	Laboratory	1.44	±0.01	3
Alate	Field	1.43	±0.01	3
Alate	Field	1.44	±0.06	10
Queen	Field	1.42	±0.02	3

closely monitored it may be difficult to non-
destructively (that is, without checking for
ovariole development) determine when an alate is
mature or whether it is an overwintering or spring
type. The specific behavior associated with
mating flights requires the correct environmental
factors as well as alates in the right
physiological state (weight, in this case, appears
to be an indicator of physiological state).

Biochemical Changes

        Insemination brings about dramatic
physiological, biochemical, and behavioral changes
that are not immediate, but occur over time. For
example, concomitant with wing muscle histolysis
(Toom et al. 1976a) and ovariole development, the
crop contents are forced into the esophagus which
expands to form a thoracic crop (Vander Meer et
al. 1982; Glancey et al. 1980). By 24 days the
majority (90 percent) of the triacylglycerols,
originally stored in the crop, are utilized.
However, hydrocarbons in the crop and esophagus
did not follow the movement of the
triacylglycerols and remained at low levels
throughout the 120 days of the experiment (Vander
Meer et al. 1982). The postpharyngeal gland opens
into the pharynx and is most highly developed in
S. invicta female sexuals (Phillips and Vinson
1980a). Hydrocarbons are the major lipid class in
the postpharyngeal gland and have been identified
as a mixture of five normal, methyl, and dimethyl
branched heptacosane hydrocarbons (Thompson et al.
1981). These compounds are species specific, are
ubiquitous in S. invicta, and can be used as
taxonomic characters (Vander Meer et al. 1985;
Vander Meer 1986a). Shortly after (about 20 days)
insemination there is a dramatic spurt of
hydrocarbon accumulation or biosynthesis in the
gland (Vander Meer et al. 1982), which decreases
to starting amounts after 120 days. Coupled with
the total quantitative changes is a qualitative
change in the hydrocarbon pattern, where the
proportion of two of the five components are
greatly reduced. The significance of these
changes is not known. Phillips and Vinson (1980b)

studied the function of the gland in S. invicta
workers and concluded that it functions as a
cephalic caecum and that the major lipid
components come from the food of the adult ant.
The postpharyngeal gland of queens, however,
probably has as yet undetermined functions.

Pheromone Production

Two S. invicta queen pheromone systems
develop as part of the dramatic physiological and
biochemical changes that occur after
insemination. A queen attractant pheromone is
stored in the poison sac of the queen and is
dispensed through the sting apparatus (Vander Meer
et al. 1980). This gives the queen control over
release of the pheromone at whatever time and
whatever quantity a particular situation
dictates. Interestingly, queen pheromone
production requires worker stimulation, since
queens removed from their colonies for more than
30 minutes are no longer attractive (Lofgren et
al. 1983). Biosynthesis of this pheromone occurs
9 to 12 days after dealation and is independent of
insemination (Glancey et al. 1981). Several
bioassays (Lofgren et al. 1983; Glancey et al.
1982) were used to isolate and identify the active
pheromone components (Rocca et al. 1983a,
1983b). Although quantitative studies on queen
attractant pheromone variation have not been
completed, it is apparent that the queen's control
of pheromone release makes the related behavioral
response highly context dependent.
A second pheromone system inhibits dealation
and oogenesis in fire ant female alates (Fletcher
and Blum 1981a). The pheromone is produced by the
colony queen and acts directly on the alates.
Since uninseminated dealates produce queen
pheromones and unfertilized eggs, they would
represent considerable competition for colony
resources. Therefore, it is to a queen's
advantage to inhibit the tendency of female
sexuals to dealate. Similar to the queen
attractant pheromone, the dealation inhibitory
pheromone is produced within 10 days of dealation
(Fletcher 1986). A bioassay for detecting this

primer pheromone was used to determine that the
queen's abdomen was the site of release (Fletcher
and Blum 1981b). In addition, the inhibitory
capability of a mated queen increases with its
fecundity (Fletcher and Blum 1983b).
Investigation of the quantitative variation of
this pheromone system awaits the isolation and
identification of the active components.

Queen fecundity, worker stimulation, queen
age, and situational context are some of the
factors that affect variation in the release of
these two important social regulatory pheromones.

NESTMATE RECOGNITION

In this section I would like to look
specifically at the behavioral and chemical
variability of recognition in fire ants although
some statements will apply to nestmate recognition
in general. Colony odor is a composite of
exogenous (environmentally derived) and endogenous
(includes genetic factors that are and are not
influenced by the environment) odors. Exogenous
odors are made up of atmospheric, soil, food, and
other odors not produced or modified by an
individual. Endogenous odors are comprised of
cuticular lipids, exocrine products, excretory
products, regurgitory products, and other odors
produced or modified by an individual and released
to the outside. This set of definitions applies
to all social insect systems and simply defines in
general terms the sources of potential recognition
cues. Odors responsible for nestmate recognition
represent a subset of colony odor and may be
composed of any possible combination of the
components that make up the exogenous and/or
endogenous components of colony odor. Examples of
the importance of exogenous (Jutsum et al. 1979,
for Acromyrmex octospinosus [Reich]) and
endogenous (Hölldobler and Wilson 1977b for
Oecophylla longinoda Latreille; Mintzer and Vinson
1985 for Pseudomyrmex ferruginea F. Smith) colony
odor recognition factors have been reported in the
literature.

Hubbard (1974) studied the influence of nest
material and colony odor on the digging behavior

of S. invicta and found that workers
preferentially dug in nest materials from their
own nests rather than that of non-nested soil or
nest material from other S. invicta colonies. He
concluded that since laboratory colony diets were
identical, ant derived odors were being
transferred to the soil. Colony odor cannot be
thought of as a unidirectional acquisition toward
colony members. An individual's exogenous and
endogenous odors can be transferred to the
environment, both passively from the cuticle and
actively from exocrine sources. For example,
species-specific cuticular hydrocarbons have been
found in S. invicta nest soil (Vander Meer,
unpublished data). There is a continual flow of
exogenous odors onto the individual's cuticle and
there is flow of both exogenous and endogenous
factors back to the environment in which the
individual resides. The flow of colony odor to
the environment may be important in territorial
recognition, and it would not be surprising if
territory and nestmate recognition cues were
identical (see Jaffé 1986). More important for
nestmate recognition is that exogenous and
endogenous odors are passed from individual to
individual through grooming and trophallaxis, so
these factors are distributed throughout a
colony. Odors that originate from the environment
but are now passed from individual to individual
are still exogenous odors. Similarly, endogenous
odors produced by an individual then passed to
others or the environment should be considered
endogenous regardless of where they are found.
The point of origin is the critical factor.
    The surface of an insect's cuticle is coated
with lipids synthesized by cells located near the
integument (Blomquist and Dillwith 1985). The
cuticle and associated lipids are ideal for the
absorption of both exogenous and endogenous
odors. It is appealing to think of the individual
ant as enveloped in chemicals that identify that
individual at both species and colony levels
(Vander Meer 1983). This follows Wilson's (1971)
statement that in all social insects, recognition
of a nestmate only involves a pause and sweep or
the antennae over the other's body. Consequently,
the search for potential nestmate recognition cues

has been directed at the analysis of cuticular
components, particularly hydrocarbons.

Cuticular Hydrocarbons

I would like to emphasize that it is not
possible to obtain a cuticular rinse that only
contains chemicals from the cuticle. Regardless
of the length of rinse time, contamination from
exocrine products, crop and postpharyngeal gland
contents, fecal material, infrabuccal pocket
pellets, etc. is always a problem with which to
contend (Vander Meer 1986a; Vander Meer,
unpublished data). These potential contaminants
must be taken into consideration whenever chemical
and/or behavioral studies are conducted on solvent
rinses or soaks of whole individuals.
Hydrocarbons are inert and chemically not
very interesting; however, they have been reported
to elicit several kinds of physiological activity
in insects; that is, sex attraction (Conner et
al., 1980), alarm (Lofquist, 1976), and parasite
attraction (Vinson et al. 1975). In addition they
have been implicated in species and caste
recognition in termites (Howard et al. 1982;
Blomquist et al. 1979). In S. invicta
hydrocarbons comprise about 75 percent of the
cuticular lipids and the five major components
have been identified as normal, methyl, and
dimethyl branched compounds (Lok et al. 1975;
Nelson et al. 1980). These hydrocarbons are
species specific and have been used as species
markers in several studies (Vander Meer 1986a;
Vander Meer et al. 1985). They are found
everywhere in the ant, from postpharyngeal gland
to the queen's ovaries (Vander Meer et al. 1982;
Vander Meer, unpublished data). Although there is
no direct evidence that S. invicta hydrocarbons
play a role in nestmate recognition, there is a
relationship between the integration of a
myrmecophilous beetle, Myrmecophodius
excavaticollis L. into fire ant colonies and the
passive acquisition of the species-specific
hydrocarbons of the host (Vander Meer and Wojcik
1982). In addition, S. richteri workers treated
with S. invicta hydrocarbons survive longer than

untreated S. richteri, when placed witn S. invicta
workers (Glancey, unpublished data). Obin (1986)
found that suppression of environmental
differences in S. invicta laboratory reared
colonies dampened but did not eliminate aggression
between workers from those colonies. He concluded
that both endogenous and exogenous cues were
important in S. invicta nestmate recognition.
Obin (1986) also found that cuticular hydrocarbon
patterns were not well correlated witn tne
measured aggressive response of laboratory
colonies on the same diet versus field colonies,
indicating that aggression at this level was keyed
to exogenous odors.

Several important questions arise when
considering hydrocarbons as potential nestmate
recognition cues for S. invicta: (1) Are
cuticular hydrocarbons a part of endogenous colony
odor? Answer: Yes. They are biosynthesized by
the ants and are released to the cuticle
(Blomquist and Dillwith 1985). (2) Are cuticular
hydrocarbons species-specific? Answer: Yes.
Analysis of four Solenopsis species in the United
States and several others from South America
demonstrated that this biochemical character is
species-specific (Vander Meer, unpublished
data). (3) Are cuticular hydrocarbons nestmate
recognition cues for S. invicta? Answer:
Unknown. There has been no direct evidence
linking cuticular hydrocarbons to S. invicta
nestmate recognition. They may or may not be
involved. (4) Are cuticular hydrocarbons readily
analyzed? Answer: Yes and no. The chemical
separation and subsequent gas chromatographic
analysis is readily accomplished, but the
differences in chromatogram patterns may not be
great and the analysis of the data usually
requires sophisticated pattern recognition
techniques (Jurs 1986). (5) Can cuticular
hydrocarbons act as a model for the fate of both
exogenous and endogenous nestmate recognition
cues? Answer: Yes. Assuming that nestmate
recognition cues are resident on the cuticle, then
whether or not hydrocarbons are involved in
nestmate recognition, they suffer (or enjoy) the
same fate as all other potential nestmate

recognition cues (both endogenous and exogenous) on the cuticle.

Hydrocarbon Pattern Variability

Recently we have analyzed the cuticular hydrocarbons of $\underline{S}$. invicta workers (reserve subcaste) by gas chromatography. Analyses of the data by several pattern recognition techniques (Jurs 1986) all clearly grouped the data according to colony. The variation within colonies was less than that between colonies. Using SIMCA pattern recognition methodology (Wold and Sjostrom 1977) a classification success rate of 100 percent was achieved. Therefore, colonies can be statistically distinguished by their cuticular hydrocarbon patterns.

Surprisingly, investigation of changes in colony hydrocarbon patterns with time (one to three month sampling intervals) showed that cuticular hydrocarbon patterns change with time. Replicates within a time period showed little variation compared to variation from time period to time period for a given colony. Hydrocarbon patterns, which at least represent what is occurring on the cuticle, are dynamically changing.

Along with changing hydrocarbon patterns, evidence for continuous mixing of colony hydrocarbons comes from a myrmecophile integration mechanism (Vander Meer and Wojcik 1982) and from the creation and chemical analysis of interspecific Solenopsis colonies (Vander Meer, unpublished data). There are at least three scenarios that explain hydrocarbon mixing and time dependent pattern changes. Scenario 1: Each individual produces a constant pattern through time but individuals differ, therefore the mean colony profile changes with time as new individuals (interindividual variation) enter the work force. Scenario 2: Individuals undergo continous random change within a specific genetic "window" of variability. Here the pattern variability is derived from both interindividual and intraindividual variation. Scenario 3: Individuals all produce the same pattern which synchronously changes with time. Scenario 3 does not require intimate mixing of hydrocarbons throughout the individuals of the colony. At this time we do not have enough information to distinguish the possibilities. If we consider that cuticular hydrocarbons at least act as a model of nestmate recognition cues, then all nestmate recognition cues found on the cuticle (exogenous and endogenous) are mixed throughout the colony, and they are dynamically changing. The cumulative evidence favors a type of Gestalt model (Crozier and Dix, 1979) for S. invicta.

Recognition Cue Imprinting

In general it is thought that there is a short (in time) critical sensitive period in a newly eclosed worker's life in which it experiences the colony's odor and imprints or learns the nestmate recognition cues (Morel 1983; Le Moli and Mori 1984). What are the implications of the above results on the learning hypothesis of nestmate recognition? (1) If colony cuticular hydrocarbon patterns change with time, then it is possible that other endogenous colony odor factors also change with time. (2) No one would argue against the statement that colony odors derived from the environment (exogenous) are dynamically changing with time. Therefore, nestmate recognition cues (a subset of colony odor) are continously changing over time, regardless of their source. Most importantly, recognition cues cannot be learned by callow workers as a fixed pattern but require continuous updating through an iterative learning process. A similar process was proposed by Wallis (1963), who worked with two Formica species. He suggested that each worker ant was probably "continually habituating to slight variations in the odour of its nestmates".

Nestmate recognition in social insects has attracted a great deal of attention, especially in the last few years. As a consequence of the inherent complexities of the potential mechanisms involved there is a plethora of terminology. The above discussion attempts to address the need to re-evaluate what has been done and evolve a more streamlined and self-consistent lexicon for

nestmate recognition. Even the term nestmate
recognition may be misleading, since the ants may
in fact be recognizing a difference in non-
nestmates rather than a similarity in nestmates.

CONCLUSION

Life within a fire ant colony is extremely
complex. Ideally, an experimenter would like to
control morphological caste, worker size, worker
age (preferentially a worker's physiological age),
colony to colony variation (we need genetically
engineered clones), colony age, the bioassay
context, the season of the year, and other
variables. Control of these variables is often
not practical or possible, but to be aware that
they exist helps to avoid some dangerous traps
associated with social insect research. This
variation makes working with fire ants both
interesting and challenging.

SUMMARY

Fire ants, especially <u>Solenopsis</u> <u>invicta</u>,
have been studied intensively for the past thirty-
five years. Variability in behavior and
biochemistry has been detected at every stage of
colony development. Nanitic workers, the first
adults produced by colony founding queens, differ
behaviorally and chemically from their mature
colony counterparts. As a colony matures the
worker size distribution changes and certain
behavior patterns are preferentially performed by
specific size categories. In addition, workers
undergo age related polyethism, moving from nurse
to forager as they become older. Venom alkaloids
(2-alkyl or alkenyl 6-methyl piperidines) vary
both qualitatively and quantitatively with the
size of the worker, but there are no differences
between temporal castes of the same size. Adult
workers require the venom alkaloids regardless of
their function; nurses use it to disinfect the
brood and nest, while foraging workers use it to
secure prey and defend the colony. The
responsiveness of workers to pheromone systems in

general diminishes with the physiological age of the individual. Also, the behavioral responsiveness of colonies often shows considerable variation, for example in initial trail formation. Female sexuals are very complicated in terms of their sexual maturity and the events, both behavioral and biochemical, that take place after mating. Nestmate recognition factors, as modeled by the cuticular hydrocarbons, have been shown to be statistically different from colony to colony and in a dynamic state of flux. In conjunction with evidence for continuous mixing of colony odor, these data suggest that individual workers continually update their perception of colony odor and, therefore, nestmate recognition cues.

REFERENCES

Adams, C.S., and Traniello, J.F.A. 1981. Chemical interference competition by Monomorium minimum (Hymenoptera: Formicidae). Oecologia 51:265-270.
Adrouny, G.A., Derbes, V.J., and Jung, R.C. 1959. Isolation of a hemolytic component of fire ant venom. Science 130:449.
Blomquist, G.J., Howard, R.W., and McDaniel, C.A. 1979. Structures of cuticular hydrocarbons of the termite Zootermopsis angusticollis (Hagen). Insect Biochemistry 9:365-370.
Blomquist, G.J., and Dillwith, J.W. 1985. Cuticular lipids. In Comprehensive insect physiology, biochemistry and pharmacology, vol. III, ed. G.A. Kerkut and L.I. Gilbert, 117-154. New York: Pergamon Press.
Blum, M.S., Walker, J.R., Callahan, P.S., and Novak, A.F. 1958. Chemical, insecticidal, and antibiotic properties of fire ant venom. Science 128:306-307.
Brand, J.M., Blum, M.S., Fales, H.M., and MacConnell, J.G. 1972. Fire ant venoms: comparative analyses of alkaloidal components. Toxicon 10:259-271.

Brand, J.M., Blum, M.S., and Barlin, M.R. 1973.
Fire ant venoms: intraspecific and
interspecific variation among castes and
individuals. Toxicon 11:325-331.
Buren, W.F. 1972. Revisionary studies on the
taxonomy of the imported fire ants. Journal
of the Georgia Entomological Society. 7:1-27.
Conner, W.E., Eisner, T., Vander Meer, R.K.,
Guerrero, A., Ghiringellie, D., and Meinwald,
J. 1980. Sex attractant of an arctiid moth
(Utetheisa ornatrix): A pulsed chemical
signal. Behavioral Ecology and Sociobiology
7:55-63.
Crozier, R.H., and Dix, M.W. 1979. Analysis of
two genetic models for the innate components
of colony odor in social Hymenoptera.
Behavioral Ecology and Sociobiology 4:217-
224.
Dumpert, K. 1981. The social biology of ants.
Boston: Pitman Advanced Publishing Program.
Fletcher, D.J.C. 1983. Three newly discovered
polygynous populations of the fire ant,
Solenopsis invicta, and their significance.
Journal of the Georgia Entomological Society
18:338-343.
Fletcher, D.J.C. 1986. Perspectives on some
queen pheromones of social insects with
special reference to the fire ant Solenopsis
invicta. In Fire ants and leaf-cutting
ants: biology and management, ed. C.S.
Lofgren and R.K. Vander Meer, 184-191.
Boulder, Colorado: Westview Press.
Fletcher, D.J.C., Blum, M.S., Whitt, T.V., and
Temple, N. 1980. Monogyny and polygyny in
the fire ant, Solenopsis invicta. Annals of
the Entomological Society of America 73:658-
661.
Fletcher, D.J.C., and Blum, M.S. 1981a.
Pheromonal control of dealation and oogenesis
in virgin queens of fire ants. Science
212:73-75.
Fletcher, D.J.C., and Blum, M.S. 1981b. A
bioassay technique for an inhibitory primer
pheromone of the fire ant Solenopsis invicta
Buren. Journal of the Georgia Entomological
Society 16:352-356.

248

Fletcher, D.J.C., and Blum, M.S. 1983a. The inhibitory pheromone of queen fire ants: effects of disinhibition on dealation and oviposition by virgin queens. Journal of Comparative Physiology A 153:467-475.

Fletcher, D.J.C., and Blum, M.S. 1983b. Regulation of queen number by workers in colonies of social insects. Science 219:312-314.

Glancey, B.M., Stringer, C.E., Craig, C.H., Bishop, P.M., and Martin, B.B. 1970. Pheromone may induce brood tending in the fire ant, Solenopsis saevissima. Nature (London) 226:863-864.

Glancey, B.M., Stringer, C.E., Jr., Craig, C.H., Bishop, P.M., and Martin, B.B. 1973. Evidence of a replete caste in the fire ant, Solenopsis invicta. Annals of the Entomological Society of America. 66:233-234.

Glancey, B.M., Stringer, C.E., Craig, C.H., and Bishop, P.M. 1975. An extraordinary case of polygyny in the red imported fire ant. Annals of the Entomological Society of America 68:922.

Glancey, B.M., Glover, A., and Lofgren, C.S. 1980. The thoracic crop in Solenopsis invicta Buren (Hymenoptera: Formicidae). Sociobiology 5:272-278.

Glancey, B.M., Glover, A., and Lofgren, C.S. 1981. Pheromone production by virgin queens of Solenopsis invicta Buren. Sociobiology 6:119-127.

Glancey, B.M., Lofgren, C.S., Rocca, J.R., and Tumlinson, J.H. 1982. Behavior of disrupted colonies of Solenopsis invicta towards queens and pheromone-treated surrogate queens placed outside the nest. Sociobiology 7:283-288.

Hölldobler, B., and Wilson, E.O. 1977a. The number of queens: an important trait in ant evolution. Naturwisensschaften 64:8-15.

Hölldobler, B. and Wilson, E.O. 1977b. Colony-specific territorial pheromone in the African weaver ant Oecophylla longinoda (Latreille). Proceedings of the National Academy of Sciences, USA 74:2072-2075.

Hölldobler, B. 1982. Interference strategy of
Iridomyrmex pruinosum (Hymenoptera:
Formicidae) during foraging. Oecologia
52:208-213.
Howard, R.W., McDaniel, C.A., Nelson, D.R.,
Blomquist, G.J., Gelbaum, L.T., and Zalkov,
L.W. 1982. Cuticular hydrocarbons of
Reticulitermes virginicus (Banks) and their
role as potential species- and caste-
recognition cues. Journal of Chemical
Ecology 8:1227-1239.
Hubbard, M.D. 1974. Influence of nest material
and colony odor on digging in the ant
Solenopsis invicta (Hymenoptera:
Formicidae). Journal of the Georgia
Entomological Society 9:127-132.
Jaffé, K. 1986. Nestmate recognition and
territorial marking in Solenopsis geminata
and in some Attini. In Fire ants and leaf-
cutting ants: biology and management, ed.
C.S. Lofgren and R.K. Vander Meer, 211-222.
Boulder, Colorado: Westview Press.
Jouvenaz, D.P., Blum, M.S., and MacConnell, J.G.
1972. Antibacterial activity of venom
alkaloids from the imported fire ant,
Solenopsis invicta Buren. Antimicrobial
Agents and Chemotherapy 2:291-293.
Jouvenaz, D.P., Lofgren, C.S., Carlson, D.A., and
Banks, W.A. 1978. Specificity of the trail
pheromones of four species of fire ants,
Solenopsis spp. Florida Entomologist 61:244.
Jurs, P.C. 1986. Pattern recognition used to
investigate multivariate data in analytical
chemistry. Science 232:1219-1224.
Jutsum, A.R., Saunders, T.S., and Cherrett, J.M.
1979. Intraspecific aggression in the leaf
cutting ant Acromyrmex octospinosus. Animal
Behaviour 27:839-844.
Le Moli, F., and Mori, A. 1984. The effect of
early experience on the development of
"aggressive" behavior in Formica lugubris
Zett. (Hymenoptera: Formicidae).
Zeitschrift für Tierpsychologie 65:241-249.
Lind, N.K. 1982. Mechanism of action of fire ant
(Solenopsis) venoms. 1. Lytic release of
histamine from mast cells. Toxicon 20:831-
840.

Lofgren, C.S., Glancey, B.M., Glover, A., Rocca, J.R., and Tumlinson, J.H. 1983. Behavior of workers of Solenopsis invicta (Hymenoptera: Formicidae) to the queen recognition pheromone: laboratory studies with an olfactometer and surrogate queens. Annals of the Entomological Society of America 76:44-50.

Lofgren, C.S., and Williams, D.F. 1984. Polygynous colonies of the red imported fire ant, Solenopsis invicta (Hymenoptera: Formicidae) in Florida. Florida Entomologist 67:484-486.

Lofquist, J. 1976. Formic acid and saturated hydrocarbons as alarm pheromones for the ant Formica rufa. Journal of Insect Physiology 22:1331-1346.

Lok, J.G., Cupp, E.W., and Blomquist, G.J. 1975. Cuticular lipids of the imported fire ants, Solenopsis invicta and richteri. Insect Biochemistry 5:821-829.

Mirenda, J.T., and Vinson, S.B. 1981. Division of labour and specification of castes in the red imported fire ant Solenopsis invicta Buren. Animal Behaviour 29:410-420.

Mintzer, A., and Vinson, S.B. 1985. Kinship and incompatability between colonies of the acacia-ant Pseudomyrmex ferruginea. Behavioral Ecology and Sociobiology 17:75-78.

Morel, L. 1983. Relation entre comportement agressif et privation sociale précoce chez les jeunes immatures de la fourmi Camponotus vagus Scop. (Hymenoptera: Formicidae). Comptes Rendus de l'Académie des Sciences, Paris, Serie III 296:449-452.

Nelson, D.R., Fatland, C.L., Howard, R.W., McDaniel, C.A., and Blomquist, G.J. 1980. Reanalysis of the cuticular methylalkanes of Solenopsis invicta and Solenopsis richteri. Insect Biochemistry 10:409-418.

Obin, M.S., and Vander Meer, R.K. 1985. Gaster flagging by fire ants (Solenopsis spp.): functional significance of venom dispersal behavior. Journal of Chemical Ecology 11:1757-1768.

Obin, M.S. 1986. Nestmate recognition cues in laboratory and field colonies of Solenopsis invicta Buren (Hymenoptera: Formicidae): effect of environment and the role of cuticular hydrocarbons. Journal of Chemical Ecology. 12:1965-1974.

Oster, G.F., and Wilson, E.O. 1978. Caste and ecology in the social insects. Princeton, New Jersey: Princeton University Press.

Phillips, S.A., Jr., and Vinson, S.B. 1980a. Comparative physiology of glands associated with the head among castes in the red imported fire ant, Solenopsis invicta Buren. Journal of the Georgia Entomological Society 15:215-226.

Phillips, S.A., Jr., and Vinson, S.B. 1980b. Source of the post-pharyngeal gland contents in the red imported fire ant, Solenopsis invicta. Annals of the Entomological Society of America 73:256-261.

Porter, S.D., and Jorgensen, C.D. 1981. Foragers of the harvester ant, Pogonomyrmex owyheei: a disposable caste? Behavioral Ecology and Sociobiology 9:247-256.

Porter, S.D., and Tschinkel, W.R. 1985a. Fire ant polymorphism (Hymenoptera: Formicidae): factors affecting worker size. Annals of the Entomological Society of America 78:381-386.

Porter, S.D., and Tschinkel, W.R. 1985b. Fire ant polymorphism: the ergonomics of brood production. Behavioral Ecology and Sociobiology 16:323-336.

Porter, S.D., and Tschinkel, W.R. 1986. Adaptive value of nanitic workers in newly founded red imported fire ant colonies (Hymenoptera: Formicidae). Annals of the Entomological Society of America. 79:723-726.

Rocca, J.R., Tumlinson, J.H., Glancey, B.M., and Lofgren, C.S. 1983a. The queen pheromone of Solenopsis invicta, preparation of (E)-6-(1-pentenyl)-2H-pyran-2-one. Tetrahedron Letters 24:1889-1892.

Rocca, J.R., Tumlinson, J.H., Glancey, B.M., and Lofgren, C.S. 1983b. Synthesis and stereochemistry of tetrahydro-3,5-dimethyl-6-(1-methylbutyl)-2H-pyran-2-one, a component of the queen recognition pheromone of Solenopsis invicta. Tetrahedron Letters 24:1893-1896.

Ross, K.G., Fletcher, D.J.C., and May, B. 1985. Enzyme polymorphisms in the fire ant, Solenopsis invicta (Hymenoptera: Formicidae). Biochemical Systematics and Ecology 13:29-33.

Ross, K.G., Vander Meer, R.K., Fletcher, D.J.C., and Vargo. E.L. 1987. Biochemical phenotypic and genetic studies of two introduced fire ants and their hybrid (Hymenoptera: Formicidae). Evolution 41:280-293.

Shorey, H.H. 1976. Animal communication by pheromones. New York: Academic Press.

Sorensen, A.A., Busch, T.M., and Vinson, S.B. 1985a. Trophallaxis by temporal subcastes in the fire ant, Solenopsis invicta, in response to honey. Physiological Entomology 10:105-111.

Sorensen, A.A., Busch, T.M., and Vinson, S.B. 1985b. Behavioral flexibility of temporal subcastes in the fire ant, Solenopsis invicta in response to food. Psyche 91:319-331.

Thompson, M.J., Glancey, B.M., Robbins, W.E., Lofgren, C.S., Dutky, S.R., Kochansky, J., Vander Meer, R.K., and Glover, A.R. 1981. Major hydrocarbons of the post-pharyngeal glands of mated queens of the red imported fire ant, Solenopsis invicta. Lipids 16:485-495.

Toom, P.M., Cupp, E.W., and Johnson, C.P. 1976a. Amino acid changes in newly inseminated queens of Solenopsis invicta. Insect Biochemistry 6:327-331.

Toom, P.M., Cupp, E.W., Johnson, C.P., and Griffin, I. 1976b. Utilization of body reserves for minim brood development by queens of the imported fire ant, Solenopsis invicta. Journal of Insect Physiology 22:217-220.

Tschinkel, W.R., and Howard, R.F. 1983. Colony founding by pleometrosis in the fire ant Solenopsis invicta. Behavioral Ecology and Sociobiology 12:103-113.

Tschinkel, W.R., and Nierenberg, N.C.E. 1983. Possible importance of relatedness in the fire ant, Solenopsis invicta Buren (Hymenoptera: Formicidae) in the United States. Annals of the Entomological Society of America 76:981-989.

Vander Meer, R.K. 1983. Semiochemicals and the red imported fire ant Solenopsis invicta Buren (Hymenoptera: Formicidae). Florida Entomologist 66:139-161.

Vander Meer, R.K. 1986a. Chemical taxonomy as a tool for separating Solenopsis spp. In Fire ants and leaf-cutting ants: biology and management, ed. C.S. Lofgren and R.K. Vander Meer, 316-326. Boulder, Colorado: Westview Press.

Vander Meer, R.K. 1986b. The trail pheromone complex of Solenopsis invicta and Solenopsis richteri. In Fire ants and leaf-cutting ants: biology and management, ed. C.S. Lofgren and R.K. Vander Meer, 201-210. Boulder, Colorado: Westview Press.

Vander Meer, R.K. 1987. Per cent emergent weight: a road map to adult rhinoceros beetle, Oryctes rhinoceros, behaviour. Journal of Insect Physiology 33:437-441.

Vander Meer, R.K., Alvarez, F., and Lofgren, C.S. Isolation of the trail recruitment pheromone of Solenopsis invicta. Journal of Chemical Ecology. In press.

Vander Meer, R.K., Glancey, B.M., Lofgren, C.S., Glover, A., Tumlinson, J.H., and Rocca, J.R. 1980. The poison sac of red imported fire ant queens: source of a pheromone attractant. Annals of the Entomological Society of America 73:609-612.

Vander Meer, R.K., Williams, F.D., and Lofgren, C.S. 1981. Hydrocarbon components of the trail pheromone of the red imported fire ant, Solenopsis invicta. Tetrahedron Letters 22:1651-1654.

254

Vander Meer, R.K., and Wojcik, D.P. 1982.
Chemical mimicry in the myrmecophilous beetle
Myrmecophodius excavaticollis. Science
218:806-808.

Vander Meer, R.K., Glancey, B.M., and Lofgren,
C.S. 1982. Biochemical changes in the crop,
oesophagus, and postpharyngeal gland of
colony-founding red imported fire ant queens
(Solenopsis invicta). Insect Biochemistry
12:123-127.

Vander Meer, R.K., Lofgren, C.S., and Alvarez,
F.M. 1985. Biochemical evidence for
hybridization in fire ants. Florida
Entomologist 68:501-506.

Vander Meer, R.K., Obin, M.S., Zawistowski, S.,
Sheehan, K.B., and Richmond, R.C. 1986. A
reevaluation of the role of cis-vaccenyl
acetate, cis-vaccenol and esterase 6 in the
regulation of mated female attractiveness in
Drosophila melanogaster. Journal of Insect
Physiology 32:681-686.

Vinson, S.B., Jones, R.L., Sonnet, P.E., Bierl,
B.A., and Beroza, M. 1975. Isolation,
identification, and synthesis of host-seeking
stimulants for Cardiochiles nigriceps, a
parasitoid of tobacco budworm. Entomologia
Experimentalis et Applicata 18:443-450.

Wallis, D.I. 1963. A comparison of the response
to aggressive behavior in two species of
ants, Formica fusca and Formica sanguinea.
Animal Behavior 11:164-171.

Wilson, E.O. 1953. Origin of the variation in
the imported fire ant. Evolution 7:262-263.

Wilson, E.O. 1962. Chemical communication among
workers of the fire ant Solenopsis saevissima
(Fr. Smith). 1. The organization of mass-
foraging. 2. An information analysis of the
odour trail. 3. The experimental induction
of social responses. Animal Behaviour
10:134-164.

Wilson, E.O. 1971. The insect societies.
Cambridge, Massachusetts: Harvard University
Press.

Wilson, E.O. 1978. Division of labor in fire
ants based on physical castes (Hymenoptera:
Formicidae: Solenopsis). Journal of the
Kansas Entomological Society 51:615-636.

Wold, S., and Sjostrom, M. 1977. SIMCA, a method
for analyzing chemical data in terms of
similarity and analogy. In Chemometrics
theory and applications, ed. B. Kowalski,
243-262. American Chemical Society Symposium
Series No. 52.
Wood, L.A., and Tschinkel, W.R. 1981.
Quantification and modification of worker
size variation in the fire ant Solenopsis
invicta. Insectes Sociaux 28:117-128.

9

Variation in Foraging Patterns of the Western Harvester Ant, *Pogonomyrmex occidentalis*, in Relation to Variation in Habitat Structure

Jennifer H. Fewell

INTRODUCTION

The ability to respond to a changing or unpredictable environment is an important adaptation because it enables an animal to exploit a wide range of environments. The social insects have successfully exploited a diversity of habitats (Wilson 1971; Brian 1978); however, the mechanisms that social insects use to adapt to unpredictable or heterogeneous environments are still not well known. Flexibility in behavioral response to environmental factors is one mechanism by which an animal can successfully adapt to an uncertain or complex environment. In this study I examine variation in foraging patterns within and among colonies of the western harvester ant, Pogonomyrmex occidentalis, in response to variation in vegetational structure. I describe a model to explain how flexibility can increase foraging efficiency at the level of the colony and at the level of the individual.

In monogynous social insects fitness is often examined at the level of colony success (Wilson 1968; Carroll and Janzen 1973; Brian 1978). The most optimal foraging strategy for an entire colony does not necessarily equal the sum of optimal foraging by individuals (Rissing and Pollock 1984). Therefore, it is important to consider the evolution of foraging patterns in social insects at the level of their effect on colony fitness. However, colony-wide foraging patterns are based on the collective behavior of individuals. Therefore, it is necessary also to consider mechanisms for colony-wide flexibility at the individual level.

Social insect colonies are often spatially fixed, and cannot migrate in response to variation in the surrounding environment. This may limit colonies to locations where resources are ample and consistently available (Brian 1978). Behavioral flexibility could allow a spatially fixed colony to continue efficient functioning in a spatially or temporally heterogeneous environment. Flexibility should be most important in contexts such as foraging that bring the colony members into direct contact with variation in the environment outside the nest.

Foraging systems in Pogonomyrmex ants have previously been characterized as species-specific (Hölldobler 1976; Davidson 1977b; Gordon 1984), although intraspecific variation in foraging pattern and food selection has been described (Bernstein 1975; Whitford 1976; Hölldobler and Lumsden 1980; Davidson 1977a). A number of studies have examined differences in foraging characteristics among species of Pogonomyrmex and their relation to species interactions in ant communities (Davidson 1977a, b; Bernstein 1979; Bernstein and Gobbel 1979; Hölldobler and Lumsden 1980; Gordon 1984). Recent studies on Pogonomyrmex and other ant genera have also begun to emphasize intraspecific interactions (Davidson 1977a; De Vita 1979) and individual variation in foraging strategy (Davidson 1977b; Traniello et al. 1984; Crawford and Rissing 1983; Rissing and Pollock 1984; Cosens and Toussaint 1985). Few studies have focused on variation in foraging behavior among colonies of a population.

Pogonomyrmex ants have two foraging patterns: (1) each individual forages independently, or (2) individuals travel on common trails, termed trunk trails, to a foraging area (Hölldobler and Wilson 1970; Hölldobler 1974). Trunk trail foraging systems are considered to be an efficient way to collect clumped food resources (Hölldobler 1976; Davidson 1977a; Whitford 1978). They may also provide smaller, more easily defended foraging areas (Hölldobler and Lumsden 1980). Individual foraging is associated with

more scattered food sources (Whitford 1978;
Bernstein 1975). Scattered resources that are
gathered individually are considered less
economical to defend territorially (Hölldobler
1974).
  Vegetational structure is a quantifiable
variable that may vary spatially and temporally.
Vegetational structure may affect foraging
efficiency by affecting the time or effort
involved in traveling to and from a resource. If
ants are able to vary foraging patterns to
increase foraging efficiency, then variation in
foraging patterns among colonies would be
predicted in environments which are heterogeneous
for vegetational structure.
  Vegetational structure would be expected to be
a more important variable in a complex habitat.
Most studies of foraging in Pogonomyrmex have been
conducted in desert environments, which are
relatively simple structurally (De Vita 1979). P.
occidentalis is found in shortgrass prairie
habitats, which are structurally more complex and
more variable than deserts. The potential for the
use of two different foraging strategies by the
ants and the location of populations in habitats
with more complex vegetation make this species a
good model for studying behavioral responses to
environmental variation.
  In this study I examine intercolonial
differences in the foraging system of P.
occidentalis and their relation to vegetation
cover surrounding the nests. I test the
hypotheses that (1) foraging strategy varies among
colonies in a population, and (2) this variation
can be predicted by the variation in vegetational
density around the nest, measured as percent
vegetational cover. I also examine longitudinal
changes in foraging strategy within colonies in
response to changes in food type availability and
distribution. With information on flexibility
both within and among colonies, I construct a
model of foraging strategy based on energetic
efficiency. This model describes how use of low
vegetational cover pathways can increase foraging
efficiency by either decreasing the energetic cost
of foraging or by increasing foraging rate.

METHODS

Fifteen colonies of Pogonomyrmex occidentalis were observed in Legion Park, an area of shortgrass prairie located 3 km east of Boulder, Colorado. The quantitative foraging data and foraging surveys used to compare colonies were collected from mid-July through mid-August, 1984. Longitudinal data were also collected on individual colonies from mid-June through September to examine changes in foraging behavior within colonies.

Vegetation in the area is common to disturbed prairie, with yucca (Yucca glauca Nutt.), prickly pear cactus (Opuntia polyacantha Haw.), and buckwheat (Eriogonum effusum Nutt.) interspersed with grasses. I categorized vegetational structure around colonies a priori, using a visual estimate of vegetational cover at ground level around each nest. The categories were: (1) nests surrounded by very low density vegetation with percent cover in most areas around the nest close to zero; (2) nests surrounded by high but uniform density vegetation, with percent cover around the nest averaging around 90 percent; (3) nests surrounded by high density vegetation interspersed with an area of very low density vegetation, often a man-made footpath. These groups will be referred to as low density (LD), high density/uniform (HD/U), and high density/variable (HD/V), respectively.

Differences in vegetational density around the nests were quantified by taking measures of vegetational cover around each nest (Table 9.1). A circle with a radius of 1 m from the nest center was drawn around each colony. Each circle was divided into twelve 30° arcs, the first arc beginning at magnetic north. I estimated the percent vegetational cover within two circular areas, 0.3 m in diameter, centered along each arc. These measures were averaged to obtain a mean vegetational cover around each nest. Variances in vegetational density were also computed and compared to demonstrate that the HD/V colonies were more variable in the amount of cover around nests than colonies of the other groups (Table 9.1).

Table 9.1
Mean vegetational cover around nests at 1 meter,
measured as percent cover

LD Group		HD/U Group		HD/V Group	
Colony	Mean cover	Colony	Mean cover	Colony	Mean cover
1	0.103	6	0.930	11	0.803
2	0.323	7	0.930	12	0.808
3	0.102	8	0.893	13	0.654
4	0.170	9	0.843	14	0.646
5	0.103	10	0.923	15	0.930
Group mean	0.160		0.903		0.768
Group SD	±0.160		±0.088		±0.263

To confirm that this population utilizes trunk
trails, I mapped the routes taken by 10 foragers
from 2 HD/V, 1 HD/U, and 1 LD colony. These
colonies were chosen because they were actively
foraging in one general direction and therefore
were most likely to have trunk trails.

I also located and mapped foraging ants from
each colony in my study group to obtain a more
complete picture of foraging activity. To
construct these foraging maps, I marked off a
series of concentric circles with radii increasing
at 0.5 m intervals from the nest center. All ants
located within 0.3 m of the circles were mapped.

Ants foraging more than 2 m from the nest were
placed on the nest area to test whether they
actually belonged to the colony being surveyed.
Ants that elicited an aggressive response from
nest ants were considered to be foreign and were
not included in the foraging map. This assay was
based on a set of preliminary experiments in which

ants from the same colony and an adjacent colony were placed on the nest surface of five colonies in the study group. These colonies were chosen because they had low nearest neighbor distances. Nest ants were watched for aggressive response. Nest ants responded aggressively to foreign ants in 86 percent of the trials (10 trials per nest). The lowest percentage for aggressive response was 70 percent. There were no aggressive responses to ants from the same colony (10 trials per nest).

Foraging activity was quantified to compare foraging patterns among nests. To obtain measures of foraging activity, a circle of radius 0.5 m centered on each nest was divided into twelve 30° arcs. Each arc was surveyed for a total of 10 minutes; the number of ants leaving across the arc and the number of ants returning with seeds across the arc were recorded. These two measures were later added to obtain a measure of total foraging activity across each 30° arc around each nest.

Surveys were conducted between 0830 and 1130 h because foraging activity was highest during this period. Foraging activity generally increased and then decreased during this time, as the temperature of the soil increased. This behavioral response to soil temperature is common among Pogonomyrmex species (Box 1960; Eddy 1966; Bernstein 1979). The timed surveys were divided into two consecutive sets of 5-minute surveys. This prevented counts in individual arcs from being biased by a change in total foraging activity in the colony over the time period of the survey.

Individual foraging patterns result in a more uniform distribution of foraging activity around a nest than does trail foraging (Hölldobler 1974; Davidson 1977b). Colonies using trunk trails will have a higher variance in foraging activity around the nest, because a greater percentage of the foragers are leaving in one or a few directions. Colonies using individual foraging patterns do not use all 360° of space around a nest equally (Hölldobler and Lumsden 1980). Therefore, distributions of foraging activity around a nest cannot be compared to a null hypothesis of a uniform distribution; however, distributions of foraging activity can be compared among colonies. The variance of the logarithm of a measure is a good statistic to compare variance in foraging activity because it is independent of the mean (Lewontin 1966). I computed the variance of the log (n + 1), where n = the number of ants observed crossing each 30° arc. The variance of log (n + 1) can be subjected to further statistical analysis (Lewontin 1966). Variance measures were compared among groups by a one-way analysis of variance.

Ants foraging from a trunk trail use the same route traveling to and from a food source. Ants foraging independently are not expected to use the same route to and from a food source, although route fidelity by individual foragers occurs (Rosengren 1971; Cosens and Toussaint 1985; Fewell, unpublished data). Individually foraging ants also use different routes from each other. Therefore, a significant correlation between the number of ants leaving across each arc and the number returning with food items across that arc is only predicted for colonies using trunk trails.

Differences in correlation coefficients among colony groups would support the hypothesis that there is a difference in the tendency to use trunk trails among ants of different colonies. I calculated Pearson product moment correlation coefficients for all colonies and compared them among groups for significance. A significant and positive correlation in the colonies already shown to be utilizing trunk trails would also demonstrate that ants counted as leaving at 0.5 m are actually leaving to forage.

To determine whether ants were traveling preferentially along low vegetation density areas, the number of ants leaving in each direction was compared to the vegetation density at 0.5 and 1 m in that direction (Pearson product moment correlation coefficient). This analysis was done for the HD/V group only, because the other groups did not have enough variation in vegetational cover around nests to test for a correlation.

A difference in food distribution around colonies could also affect foraging patterns. The HD/U and HD/V colonies were interspersed in the study area and were likely to have access to

similar food species. The majority of the LD
colonies were located in a different area of the
study site. These colonies could be collecting
different food species with different distribution
patterns. If a colony is collecting a clumped
resource, a high proportion of items collected
will be of one species. Foraged items were
collected from returning foragers at intervals
throughout the season for six colonies (two from
each group); this enabled me to make diversity
comparisons among colonies and longitudinally
within colonies. All objects were taken from
foragers returning to the nest for a period of one
hour. Diversity indices (D) were calculated for
each collected set, using the equation

$$D = -\Sigma_i (P_i \log P_i),$$

where P equals the proportion of a species found
in sample i (Webb 1974). Mean diversity indices
and species richness of foraged items were
compared among nests and longitudinally for
individual nests.

RESULTS

The foraging pattern exhibited by P.
occidentalis varied with the vegetational
structure around the nest site. All five colonies
surrounded by variable density vegetation (HD/V
colonies) had major trunk trails (see foraging
maps, Figure 9.1). Over 60 percent of the mapped
ants foraging from these colonies were utilizing
trunk trails.
Trunk trails were less extensive in the
colonies surrounded by even vegetation. One HD/U
colony (Colony 6) also had a large trunk trail;
three had smaller trails, involving fewer than 30
percent of foraging ants, and one had no obvious
trunk trails (Colony 10). Distribution of
foragers from the LD colonies was also scattered;
no obvious trunk trails appeared on the foraging
maps.

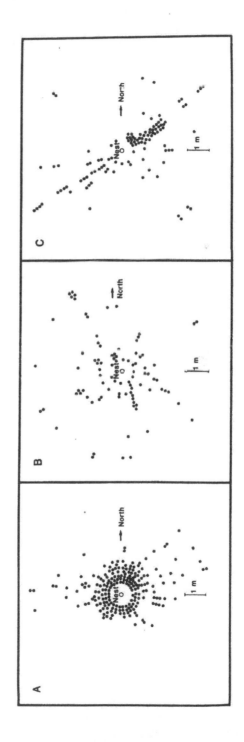

Figure 9.1. Survey of foragers from (A) LD Group Colony 3, (B) HD/U Group Colony 8, (C) HD/V Group Colony 11.

Ants using trunk trails generally follow a specific route, which is marked or identified in some way (Hölldobler 1974, 1976; Hölldobler and Wilson 1970). In a number of colonies I observed large numbers (often over 100) of ants following common routes to forage. I confirmed the ability of ants to follow a specific route by mapping the routes of individual foragers (Figure 9.2). Five of 10 ants from HD/V Colony 11 used the same route to forage, 4 used a second route, and 1 foraged independently of the other 9. The 2 common routes were followed exactly, with some divergence at the actual foraging area. Similar fidelity to a route was observed in another HD/V colony (Colony 12). Marked foragers showed less fidelity to common routes in the HD/U and LD colonies (Figure 9.2).

These observational data are supported by the quantitative measures of variability in foraging activity around the nest. I predicted that colonies using trunk trails would have a greater variance in foraging activity around the nest than colonies whose workers forage independently. Foraging activity around the nest had significantly higher variance in the HD/V group than in the other two groups (one-way ANOVA, $P <$ .001, with Duncan's a posteriori test for homogeneous subsets, $P <$ .05; the HD/U and LD group colonies had more even distributions of foraging activity around the nest than did HD/V colonies. The variance of the log of foraging activity around nests did not differ significantly between the HD/U and LD groups (Table 9.2).

Colonies using trunk trails should have stronger correlations between the number of foragers leaving across a given arc and the number returning with seeds across the same arc than colonies whose workers are foraging independently. The correlation between the number of ants leaving and the number returning with seeds for a given direction was highly significant in all 5 HD/V colonies (Table 9.3). The r-value for the HD/V colonies was much greater than for either of the other 2 groups (Table 9.3); however, significant correlations were also seen for 3 of the other 10 colonies. One of these colonies (Colony 6) did show evidence of a major trunk trail in the foraging maps.

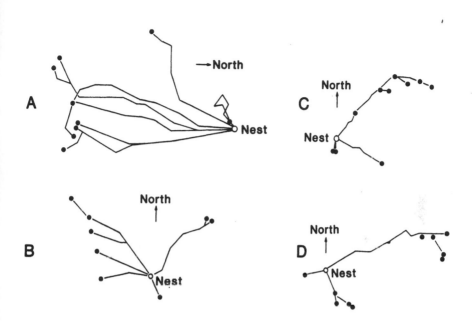

Figure 9.2. Routes taken by individual foragers
    from (A) LD Group Colony 4, (B) HD/U Group
    Colony 8, (C) HD/V Group Colony 11, (D) HD/V
    Group Colony 12.

    The large trunk trails of the HD/V colonies
usually led through areas of low density
vegetation. For 4 of the 5 HD/V colonies, a
statistically significant negative correlation was
found between foraging activity across each 30°
arc around the nest and the percent cover of
vegetation centered on that arc (Table 9.4). Ants
tend to leave the nest in directions with low
density vegetation.
    Ants from Colony 15, placed a priori into the
HD/V group, used a small vole pathway for the
route of their largest trunk trail. Although
these ants were using a route of extremely low
vegetational cover, vegetation density analysis
was not fine enough to detect the vole trail, and
the measured vegetation density did not vary
around that nest.
    Foraging activity at 0.5 m and vegetational
density at 1 m were also negatively correlated
(Table 9.4), demonstrating that the tendency to

Table 9.2
Comparison of the variance of the log of foraging
activity around nest among colony groups[a]

	Leaving	Returning with Items	Total
F-ratio	25.850	26.685	35.246
Significance Level (P)	< .001	< .001	< .001
Homogeneous Subsets[b]			
Subset 1	HD/U, LD	HD/U, LD	HD/U, LD
Subset 2	HD/V	HD/V	HD/V

[a]Analysis is performed using one-way ANOVA.
[b]Differences among groups obtained using Duncan's
a posteriori division into homogeneous (P > 0.5)
subsets.

choose a path of low-density vegetation extends
well beyond the nest site. Nest 13 did not show a
significant correlation between foraging at 0.5 m
and vegetation density at 1 m. The path of low
density vegetation and the trunk trail from this
nest made a right angle turn at 0.75 m.

Mean diversity indices and mean species
richness of foraged items were compared among the
three groups to determine whether clumping of
resources could explain the differences in
foraging pattern among groups. Data were pooled
for the entire study period (June to September)
and also separately for the period that
quantitative foraging data were collected (29 June
to 11 August). Mean species richness and mean
diversity indices of the three groups were very
similar (Table 9.5). At no time did the HD/V
group show a lower mean species richness or mean
diversity of species collected than the other two
groups. This analysis indicates that the tendency
for HD/V colonies to use trunk trails is not due

Table 9.3
Correlation between number of ants leaving from
nest sectors and number returning with foraged
items

LD Group		HD/U Group		HD/V Group	
Colony	r^a	Colony	r^a	Colony	r^a
1	0.1174	6	0.8173^d	11	0.9875^d
2	0.3937	7	0.4640	12	0.9827^d
3	0.7404^c	8	0.1998	13	0.8294^d
4	0.1614	9	-0.0856	14	0.9410^d
5	0.4917	10	-0.2204	15	0.7979^d
Pooled data	0.4331^d		0.2856^b		0.9155^d

[a]Pearson product moment correlation coefficient
values.
[b]Significant at $\underline{P}$ < .05.
[c]Significant at $\underline{P}$ < .01.
[d]Significant at $\underline{P}$ < .001.

to their specialization on fewer resource types
than the other colonies.
    Some colonies showed temporal changes in
foraging pattern in response to clumped
resources. Ephemeral trail systems were observed
in some colonies. These were generally not
quantified. Such systems included recruitment to
a dead grasshopper from Colony 10, and a longer
lived trunk trail to a small gravel pile by Colony
9. Both of these colonies were in the HD/U group.
    Colony 4, an LD colony, foraged evenly around
the nest through most of the summer. However, the
foraging pattern changed in September when a patch
of Echinochloa crus-galli (L.) P. Beauv. (barnyard
grass) came into bloom. Eighty percent of the
items collected by Colony 4 in a September survey
were seeds of this species. During the period

Table 9.4
Correlation between total foraging activity and
vegetational cover in 30° sectors around nests[a]

Colony	Vegetation measured 0.5 m from nest	Vegetation measured 1 m from nest
11	-0.8802^d	-0.6636^c
12	-0.5822^b	-0.8354^d
13	-0.8945^d	-0.4270
14	-0.5382^b	-0.7607^c
Pooled data	-0.5011^d	-0.5043^d

[a]Pearson product moment correlation coefficients.
[b]Significant at $\underline{P} < .05$.
[c]Significant at $\underline{P} < .01$.
[d]Significant at $\underline{P} < .001$.

that Echinochloa was being collected the majority
of the foragers from Colony 4 left the nest across
one 30° arc (routes of individual foragers shown
in Figure 9.2A). However, foraging ants did not
utilize a trunk trail to gather these seeds.
    Colony 12, an HD/V colony, switched the
direction of its major trunk trail through the
summer. During foraging surveys made in July and
August approximately 50 percent of all ants
traveled along the major trunk leaving the nest
across the arc between 30° and 60° northeast. In
one survey of food item diversity made after
Eriogonum effusum (bushy buckwheat) began to bloom
in September over 99 percent of the ants from
Colony 12 observed with food items had seeds from
this plant. This colony showed a corresponding
change in diversity of foraged items from a mean
index of 0.645 from June through August to 0.192
in September. The trunk trail seen in July and
August was abandoned at this time. Seventy-three
percent of foraging ants were located on a trunk
trail to and from Eriogonum bushes located between
0° and 30° from the nest. Another 14 percent were

Table 9.5
Mean diversity index (D) and mean species richness
(R) for foraged items

	29 June- 11 Aug.	6 Sept.- 20 Sept.	Total Season
		LD Group	
Number of nests	2	3	5
R	5.5	7.0	6.5
D	0.441	0.506	0.474
		HD/U Group	
Number of nests	3	3	6
R	6.7	5.3	6.0
D	0.718	0.702	0.710
		HD/V Group	
Number of nests	4	5	9
R	6.0	6.2	6.1
D	0.674	0.553	0.607

on a trunk trail between 240° and 270° to an
Eriogonum bush northwest of the nest. These
results demonstrate that changes in foraging
patterns occur within P. occidentalis colonies.

DISCUSSION

   P. occidentalis demonstrates flexible
foraging patterns in response to environmental

variation. Variation in foraging pattern among
colonies has been reported in other species of
Pogonomyrmex, but not measured quantitatively
(Whitford 1976, 1978; Hölldobler 1974, 1976;
Davidson 1977b; Gordon 1984). The agreement
between the quantitative distributions of foraging
activity and the more traditional observational
methods of this study indicates that the
quantitative measures used provide a useful,
statistically treatable method to compare
intercolonial differences in foraging pattern.

Foraging patterns differ among colonies in
direct relation with the vegetational structure
around nests. Colonies in areas with cleared
pathways rely more heavily on trunk trails for
food gathering. The ants from these colonies
position trails through areas of lowest
vegetational cover around the nest.

Individual colonies also demonstrate
flexibility in foraging patterns. The intensity
of trunk trail use and the location of trails are
varied in response to changes in the distribution
of preferred resources. Certain clumped resources
stimulated the use of trunk trails by colonies
that normally exhibited individual foraging. The
appearance of Eriogonum effusum flowers resulted
in dramatic changes in the foraging patterns of a
number of colonies. This plant represents a
clumped, highly preferred resource. Rogers (1972)
noted that E. effusum accounted for 27 percent of
the seeds gathered by his study population of P.
occidentalis. Other clumped resources also
resulted in smaller trunk trails in all 3 colony
groups.

LD Colony 4 used an individual trail system
to gather Echinochloa crus-galli, a highly
preferred, but evenly distributed resource. The
lack of a trunk trail system to this resource
suggests, anecdotally, that ants respond to the
clumped distribution of the resource by using a
trail system.

Previous studies have focused on the
importance of trunk trails in the defense of
resources (Hölldobler 1974, 1976; De Vita 1979;
Hölldobler and Lumsden 1980; Harrison and Gentry
1981). Colonies in my study population did not
defend trunk trails aggressively near foraging

areas. Similar observations have been made for a nearby population of this species (Fewell, unpublished data). It is possible that resources are present at a sufficiently high level in this habitat that competition is relatively low between neighboring colonies. Under these circumstances trunk trails may exist simply to take advantage of the energetic benefits of foraging over cleared areas.

Foraging patterns in ants are theoretically related to foraging efficiency and energetic cost (Whitford 1978; Traniello et al. 1984). The demonstration that colony foraging patterns can be influenced directly by vegetational structure suggests that colonies are working to maximize foraging efficiency. The efficiency of a particular foraging pattern is affected by the foraging rate that individual ants might achieve and by the energetic efficiency of the particular foraging pattern. Dense vegetation increases the three-dimensional space through which an ant has to travel to and from a food source and might affect foraging rate or effort.

Whether this increase in distance has a significant energetic cost to the colony is difficult to evaluate. Jensen and Holm-Jensen (1980) reported a cost of running of 0.44 μl CO_2/m in **Formica** **rufa** L., which is close to the size of **Pogonomyrmex** **occidentalis**. Assuming a respiratory quotient of 0.8 (from Jensen and Holm-Jensen 1980) and a Q_{10} temperature effect of 2, it would cost an ant 5×10^{-2} calories to travel 10 meters at 30° C. Ten meters would be a reasonable round trip distance from the nest for a forager on a trunk trail.

Kelrick and MacMahon (1985) and Kelrick et al. (1986) have shown that **Pogonomyrmex** **occidentalis** workers selectively choose seeds with higher soluble carbohydrate content, suggesting that they are actually selecting seeds based on some measure of energy content. Total caloric value per seed in their sample was variable, but averaged 33.37 calories per seed (Kelrick and MacMahon 1985). The fraction of these calories that is potentially assimilated may be roughly estimated by the fraction of the seed lacking cellulose.   This averaged 67 percent of total seed weight.

The estimated cost of a foraging trip is 2 orders of magnitude lower than the estimated energetic value of a seed.   It is possible that energetic cost is not important to the individual forager when costs are this much below benefits. However, energetic cost will be increased by additional factors which cannot be estimated from the given data.   The energetic cost of traveling would roughly double at 40° C, which is close to the preferred foraging temperature for this species (Eddy 1966; Fewell, unpublished data). The cost will also increase when the ant is carrying a load.   The largest load carried by an ant in this study weighed 95.4 mg.   Kelrick et al. (1986) report seeds weighing approximately 80 mg carried by this species.   Cost increases linearly with increasing load in vertebrates (Taylor et al. 1980).   Therefore, an ant carrying an 80 mg seed at 40° C would have an approximate energy expenditure of 0.4 cal/trip.

The cost of traveling a given distance is also increased if the animal travels at a low speed.   An ant moving through dense cover travels at a noticeably slower rate than one on a smooth direct route, suggesting that travel through vegetation further increases both time and energetic cost per trip.   Finally, not all foraging trips are successful.   If only 1 in 3 trips is successful, the energetic cost per trip could approach 10 percent of the average benefit per seed.   While these estimates strongly suggest that energetic cost is likely to be low relative to the caloric value of seeds, it is clear that actual measurements of the parameters estimated would be extremely useful.

Data from other studies indicate that energetic cost is an important variable in foraging decisions.   Individual Formica decrease selectivity in prey choice at higher body temperatures (Traniello et al. 1984).   Traniello et al. suggested that this was due to a higher metabolic cost of foraging at higher body temperature.   They proposed that foragers minimize the cost of trips by decreasing selectivity at higher temperatures.   On the other hand, Davidson

(1978) reports increased selectivity with
increasing foraging distance in Pogonomyrmex.
These studies suggest that energy expenditure to
gain a food source affects forager selectivity.

Increasing the three-dimensional distance to
a food source would also affect foraging time and
therefore the foraging rate achievable by
individual workers. Running uphill is
energetically less costly to small vertebrates
than to large vertebrates (Taylor 1972), so that
the increased angle of travel may not increase the
energetic cost of travel, but it seems likely to
affect the time necessary to travel a given
distance measured horizontally from the nest.
This could decrease the number of foraging trips
made per day and decrease the amount of food
brought back to the nest, potentially decreasing
the production of brood (MacKay 1981).

If either energetic cost per foraging trip or
time required to walk a given distance is affected
significantly by the density of vegetation through
which the forager moves, then areas of low
vegetational cover would enable ants to travel
farther from the nest. This would increase the
total area that could be covered by the foragers
of a colony within a unit time. In Figure 9.3A
the foraging area of a colony is drawn, assuming
that ant walking speed in clear ground is twice
the speed in grassy areas. In Figure 9.3B it is
assumed that ants may walk four times faster on
clear ground. The model holds if food is either
randomly or uniformly distributed.

Three predictions can be drawn from this
model. First, ants would be expected to forage
farther in areas of low vegetational density.
Ants from LD colonies or HD/V colonies forage
approximately twice as far as ants foraging in
grass, suggesting that this prediction will be
fulfilled.

Second, in areas of evenly or randomly
distributed resources, foraging distance from the
nest should increase along a pathway. Ants will
travel into the vegetation along the pathway;
however, the distance ants travel into vegetation
along the path should decrease with distance
traveled along the path (Figure 9.3). The third
prediction of the model is that either energetic

cost or traveling time for an ant to move a unit
distance over cleared ground will be less than
over grassy ground.

The cleared pathway provides a common route
for a large number of workers traveling to and
from food sources. Use of a pheromonal trunk
trail system would benefit workers returning to
food sources along such a path by providing a
clearly marked route. Trunk trails on a low
vegetational cover pathway might be more easily
marked and identified than through the more

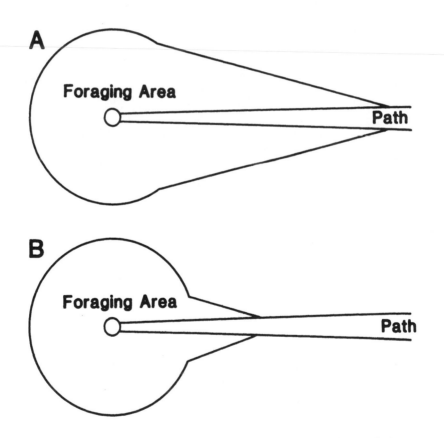

Figure 9.3.   Predicted foraging area when (A)
       walking speed is four times faster or
       energetic cost of traveling is 1/4 as much on
       the path, and when (B) walking speed is two
       times faster or energetic cost of traveling
       is 1/2 as much on the path.

complex higher vegetation, enabling ants to travel faster. Such benefits to the colony might outweigh those from an individual ant foraging randomly, although food encounter rates of an individual ant going through vegetation might be higher than for an ant using a trunk trail. Therefore, the behavior that optimizes foraging success for an individual may not be the behavior that optimizes colony foraging success (Rissing and Pollock 1984).

A highly preferred clumped resource would have a significant effect on a system based on efficiency and energetic return. A flexible foraging system would allow the colony to switch foraging patterns in response to the appearance of a preferred food source. Pheromonal trunk trails would be advantageous in guiding foragers to this resource, facilitating its exploitation (Hölldobler 1974).

The mechanisms for switching in response to such variables must occur at the level of the individual. It is unknown which subset of the foraging population will respond to a new resource. Data collected on one colony indicate very high route fidelity by individual Pogonomyrmex occidentalis workers on trunk trails. No ants observed on trunk trails were observed to forage individually. Some individually foraging ants later switched to trunk trails, while others were observed to continue to forage individually for one week (Fewell, unpublished data). These preliminary results suggest that there is specialization of foraging mode, and that, at least in this colony, switching from trunk trail use to the individual foraging mode is less likely. However, the abandoning of one major trunk trail and the developing of another by Colony 12 suggest that ants using a trunk trail are capable of changing their behavior when a highly preferred resource becomes available.

The work of Davidson (1978), Crawford and Rissing (1983), Kelrick et al. (1986), Breed et al. (1987) and others demonstrates that ants have the ability to evaluate the relative rewards of different food resources. It is possible that information on resource quality and accessibility is used in switching decisions. In such a system, routes through cleared vegetation would be maintained by continued route fidelity of individual foragers. As long as resource quality in these areas was relatively similar to that on other routes, it would be economical to continue to forage on these trails. Foragers going to equally preferred resources over more energetically demanding routes would be more easily recruited to new resources. Ants using trunk trails will switch if a new resource of higher net quality is found.

In conclusion, P. occidentalis foraging patterns are flexible both among colonies of a population and within colonies observed over time. This flexibility occurs in response to the immediate environmental variables of variation in vegetational density and clumping of preferred resources. Flexibility in colonial foraging patterns may lead to enhanced foraging efficiency and colony fitness.

SUMMARY

Ants may vary foraging patterns to increase foraging efficiency in complex or changing environments. In particular, ant foraging patterns may be influenced by vegetational structure, which can vary spatially and temporally. Foraging patterns were compared among and within colonies of the western harvester ant, Pogonomyrmex occidentalis (Cresson), which were located in areas of: (1) high, even vegetational cover; (2) low, even vegetational cover; (3) high vegetational cover with low vegetational cover pathways through it. The tendency of a colony's foragers to use trunk trails versus foraging independently was measured quantitatively by comparing the variance in foraging activity around nests. The quantitative data were supported by cartographic surveys of foraging activity around nests and by observations of individual foragers. Ants from colonies surrounded by even vegetational cover tended to forage independently. Ants from colonies surrounded by variable vegetational cover tended to travel on trunk trails located along the low vegetational cover pathways. Use of low vegetational cover pathways may decrease the amount of effort or time a forager spends traveling a unit distance, increasing foraging efficiency. Ants also responded to clumped resources by forming trunk trails. Foraging patterns in this species vary in response to both vegetational structure and clumping of resources.

ACKNOWLEDGMENTS

This work was supported in part by awards from the Kathy Lichty Memorial Fund and the University of Colorado Graduate School Foundation Fund. I thank J. Harrison, M. Breed and K. Williamson for advice and comments on the manuscript.

REFERENCES

Bernstein, R.A. 1975. Foraging strategies of ants in response to variable food density. Ecology 56:213-219.

Bernstein, R.A. 1979. Relations between species diversity and diet in communities of ants. Insectes Sociaux 26:313-321.

Bernstein, R.A., and Gobbel, M. 1979. Partitioning of space in communities of ants. Journal of Animal Ecology 48:931-942.

Box, T.W. 1960. Notes on the harvester ant, Pogonomyrmex barbatus var. molefacieus, in South Texas. Ecology 41:381-382.

Breed, M.D., Fewell, J.H., Moore, A.J., and Williams, K.R. 1987. Graded recruitment in a ponerine ant. Behavioral Ecology and Sociobiology 20:407-411.

Brian, M.V., ed. 1978. Production ecology of ants and termites. New York: Cambridge University Press.

Carroll, C.R., and Janzen, D.H. 1973. Ecology of foraging by ants. Annual Review of Ecology and Systematics 4:231-257.

Cosens, D., and Toussaint, N. 1985. An experimental study of the foraging strategy of the wood ant Formica aquilonia. Animal Behaviour 33:541-552.

Crawford, D.L., and Rissing, S.W. 1983. Regulation of recruitment by individual scouts in Formica oreas Wheeler (Hymenoptera, Formicidae). Insectes Sociaux 30:177-183.

Davidson, D.W. 1977a. Species diversity and community organization in desert seed-eating ants. Ecology 58:711-724.

Davidson, D.W. 1977b. Foraging ecology and community organization in desert seed-eating ants. Ecology 58:725-737.

Davidson, D.W. 1978. Experimental tests of the optimal diet in two social insects. Behavioral Ecology and Sociobiology 4:35-41.

De Vita, J. 1979. Mechanisms of interference and foraging among colonies of the harvester ant Pogonomyrmex californicus in the Mojave desert. Ecology 60:729-731.

Eddy, T.A. 1970. Foraging behavior of the western harvester ant, Pogonomyrmex occidentalis, (Hymenoptera: Formicidae) in Kansas. Ph.D. diss., Kansas State University, Manhattan.

Gordon, D.M. 1984. Species-specific patterns in the social activities of harvester ant colonies (Pogonomyrmex). Insectes Sociaux 31:74-86.

Harrison, J.S., and Gentry, J.B. 1981. Foraging patterns, colony distribution and foraging range of the Florida harvester ant Pogonomyrmex badius. Ecology 62:1467-1473.

Hölldobler, B. 1974. Home range orientation and territoriality in harvesting ants. Proceedings of the National Academy of Sciences, USA 71:3271-3277.

Hölldobler, B. 1976. Recruitment behavior, home range orientation, and territoriality in harvester ants, Pogonomyrmex. Behavioral Ecology and Sociobiology 1:3-44.

Hölldobler, B., and Lumsden, C.J. 1980. Territorial strategies in ants. Science 210:732-739.

Hölldobler, B., and Wilson, E.O. 1970.
Recruitment trails in the harvester ant
Pogonomyrmex badius. Psyche 77:385-399.
Jensen, T.F., and Holm-Jensen, I. 1980.
Energetic cost of running in workers of three
ant species, Formica fusca L., Formica rufa
L., and Camponotus herculeanus L.
(Hymenoptera, Formicidae). Journal of
Comparative Physiology 137:151-156.
Kelrick, M.I., and MacMahon, J.A. 1985.
Nutritional and physical attributes of seeds
of some common sagebrush-steppe plants: some
implications for ecological theory and
management. Journal of Range Management
38:65-69.
Kelrick, M.I., MacMahon, J.A., Parmenter, R.R.,
and Sisson, D.V. 1986. Native seed
preferences of shrub-steppe rodents, birds
and ants: the relationship of seed attributes
and seed use. Oecologia 68:327-337.
Lewontin, R.C. 1966. On the measurement of
relative variability. Systematic Zoology
15:141-142.
Mackay, W.P. 1986. A comparison of the
ecological energetics of three species of
Pogonomyrmex harvester ants (Hymenoptera:
Formicidae). Ph.D. diss., University of
California, Riverside.
Rissing, S.W., and Pollock, G.B. 1984. Worker
size variability and foraging efficiency in
Veromessor pergandei (Hymenoptera:
Formicidae). Behavioral Ecology and
Sociobiology 15:121-126.
Rogers, L.E. 1972. The ecological effects of the
western harvester ant (Pogonomyrmex
occidentalis) in the shortgrass plains
ecosystem. United States International
Biological Program Technical Report 206:1-
109.
Rosengren, R. 1971. Route fidelity, visual
memory and recruitment behavior in foraging
wood ants of the genus Formica (Hymenoptera,
Formicidae). Acta Zoologica Fennica 133:1-
106.
Taylor, C.R. 1972. Running up and down hills:
some consequences of size. Science 178:1096-
1097.

Taylor, C.R., Heglund, N.C., McMahon, T.A., and
    Looney, T.A. 1980. Energetic cost of
    generating muscular force during running: a
    comparison of large and small animals.
    Journal of Experimental Biology 86:109-118.
Traniello, J.F.A., Fujita, M.S., and Bowen, R.V.
    1984. Ant foraging behavior: ambient
    temperature influences prey selection.
    Behavioral Ecology and Sociobiology 15:65-68.
Webb, D.J. 1974. The statistics of relative
    abundance and diversity. Journal of
    Theoretical Biology 43:277-291.
Whitford, W.G. 1976. Foraging behavior of
    Chihuahuan Desert harvester ants. American
    Midland Naturalist 95:455-458.
Whitford, W.G. 1978. Foraging in seed-harvester
    ants Pogonomyrmex spp. Ecology 59:185-189.
Wilson, E.O. 1968. The ergonomics of caste in
    the social insects. American Naturalist
    102:41-66.
Wilson, E.O. 1971. The insect societies.
    Cambridge, Massachusetts: Harvard University
    Press.

10

Variation in Behavior Among Workers of the Primitively Social Wasp *Polistes fuscatus variatus*

David C. Post, Robert L. Jeanne,
and Eric H. Erickson, Jr.

INTRODUCTION

Because it represents a primitive stage in the evolution of the complex social organization seen in more advanced species, Polistes is an important genus for the study of the evolution of social behavior. Polistes colonies are relatively small, reproductive dominance apparently is maintained by physical aggression, physical castes are lacking, and division of labor is incomplete, with queens often performing worker-like tasks (for example, foraging) (for a review see Jeanne 1980). Previous studies of the social organization of Polistes societies have elucidated the differences in behavior between the queen and worker castes. Queens lay most of the eggs, spend more time on the nest, initiate most of the cells, forage primarily for nesting materials, and perform behavior generally associated with dominance (for example, abdomen wagging and lateral vibrations); workers do most of the foraging for food and spend relatively little time on the nest (Pardi 1948, 1951; Owen 1962; Yoshikawa 1963; West Eberhard 1969; Dew and Michener 1981; Gamboa and Dew 1981; Dew 1983; Kasuya 1983).

Most studies of division of labor in social wasps have concentrated on clarifying the differences in behavior between the worker and queen castes. Yet previous studies show that there are differences in the behavior among individual workers. Workers differ primarily in their foraging activity (Owen 1961; Yoshikawa 1963; Dew and Michener 1981; Strassmann et al. 1984). Workers emerging early in the season tend to forage more than those emerging later and some of the foragers specialize on certain items only (for example, prey). Strassmann et al. (1984) found that workers of P. exclamans Viereck can be classified into two categories on the basis of their foraging activity: foragers for prey (caterpillars) and non-foragers. In a laboratory study of the foraging behavior of workers of P. metricus (Say), Dew and Michener (1981) found that foraging behavior was influenced by the foraging activities of nestmates. The oldest worker on each nest usually was the primary forager, while later emerging workers rarely foraged unless the primary forager was removed from the nest.

In the present study we quantified the behavior of individual workers of Polistes fuscatus variatus Cresson on the nest during the ergonomic stage of colony development (as defined by Oster and Wilson [1978]). First, we provide a complete descriptive catalog of behavioral acts by workers and the relative frequency of occurrence of each type of act. Second, we analyze differences in the lifetime behavioral repertories among workers with emphasis on behavior associated with pulp and prey handling. Third, we describe age-related changes in behavior and individual variation in age polyethism. Finally, we look at individual differences in the frequency of a returning forager's giving the entire load of pulp or prey to a nestmate to utilize on the nest (task partitioning [Jeanne 1986]) to determine if some workers exhibit task partitioning, while others do not.

MATERIALS AND METHODS

Data Collection

Behavioral observations were conducted from 28 June to 18 August 1981 on 7 colonies located in small (20 x 20 x 30 cm) wooden nesting shelters at the University of Wisconsin Rieder Experimental

Farm in Madison. Colonies 3 and 7 were initiated in the nesting shelters by lone foundresses (queens) during the second week of May. Colonies 11, 13, 14, 15, and 16 were collected around Dane County, Wisconsin, and transplanted without anaesthesia to nest boxes during the first week of June, at least three weeks prior to worker emergence. All of the colonies remained queen-right throughout the duration of the study.

Worker wasps were marked for individual recognition with Testor's PLA® enamel. All workers that emerged on Colonies 3 (N = 9 workers), 7 (N = 6), 14 (N = 5), 15 (N = 9), and 16 (N = 5) were observed. Colonies 11 and 13 produced 13 and 17 workers, respectively, within ten days after the first worker emerged. During this period only the first five workers that emerged on each of these colonies were marked and observed. After the first ten days all newly emerged workers were marked and observed in both nests (N = 7 for Colony 11 and N = 3 for Colony 13). The workers in each colony were numbered according to their order of emergence on the colony. The date of emergence of the first worker on each nest was as follows: Colonies 3: 7 July; 7: 4 July; 11: 2 July; 13: 28 June; 14: 5 July; 15: 6 July; 16: 8 July. Workers (N = 54) were observed for a total of 537.5 female-hours. Colonies 3 and 13 were singled out for detailed analysis, since most workers on these survived to the end of the summer (Table 10.1), providing a large number of events recorded from these colonies and thus facilitating comparison among workers.

To determine the frequency of acts performed by workers, we used the "focal animal sampling technique" (Lehner 1979), in which all acts performed by marked individuals were recorded over a 30-minute period. All of the acts were defined such that no act followed itself (see Results). Acts were recorded sequentially on paper using one- or two-letter codes for each type of behavioral act and then later transcribed into a computer data file for analysis. Since workers spent much of their time off the nest, inactive on the nest, or engaged in behavioral acts of long duration (for example, nest construction and

Table 10.1
Life histories and observation data for the workers in Colonies 3 and 13[a]

Worker Number	Date Emerged	Last Observed	Number of Days Observed	Number of Hours Observed	Percent time off nest
			Colony 3		
1	7 July	24 July	18	11.0	61.5
2	10 July	18 Aug	40	19.5	56.6
3	11 July	17 Aug	39	18.5	56.5
4	12 July	17 Aug	38	16.0	72.1
5	16 July	17 Aug	34	17.5	60.5
6	16 July	18 Aug	34	17.0	34.3
7	17 July	18 Aug	33	14.5	68.8
8	20 July	18 Aug	30	12.0	30.0
9	21 July	18 Aug	29	14.0	60.0

Colony 13

1	28 June	18 Aug	52	26.0	62.6
2	30 June	18 Aug	50	27.0	69.5
3	1 July	17 Aug	48	25.0	56.5
4	2 July	4 Aug	34	21.0	67.5
5	2 July	18 Aug	48	25.0	51.4
18	23 July	18 Aug	27	7.5	64.5
19	27 July	18 Aug	23	7.0	1.0
20	29 July	18 Aug	21	7.0	0

[a]Wasps last observed on 18 August survived until the last day of observations.

grooming), up to five individuals could be observed simultaneously. Observations were conducted continuously between 0930 and 1730 h while the wasps were active, unless interrupted by inclement weather. An attempt was made to observe each colony for one hour each day. The order in which the colonies were observed each day was randomized to insure that each colony was observed at various times during the day. Observations continued until the first male emerged on each colony.

Data Analysis

Information theory was used to measure differences between the behavioral repertories of different workers and between the behavioral repertories of workers of different ages. Information theory provided a measure of (1) the divergence (D_1) from equal probability (H_{MAX}) of rows (behavioral acts) and columns (individuals or age categories) in a matrix, and (2) the divergence (D_2) from independence of the behavioral repertory and individual or age (H_{2MAX}) (Baylis 1976; Kolmes 1985). For example, a high D_1 relative to H_{MAX} for behavioral acts indicated that the behavioral patterns recorded were not performed at similar frequencies. A high D_2 relative to H_{2MAX} indicated that the behavioral acts and the individual or age category were not independent, thus providing evidence of worker or age specialization.

D_1 was calculated using the following formulas:

$$H_{1MAX} = \log_2 N,$$

$$H_1 = - \Sigma P_i \log_2 P_i,$$

$$D_1 = H_{1MAX} - H_1,$$

where N is the number of rows or columns and P_i is the probability of the sum of the ith row or column.

D_2 was calculated as follows:

$$H_{2MAX} = H_{1ROWS} + H_{1COLUMNS}$$

$$H_2 = - \Sigma P_{ij} \log_2 P_{ij}$$

$$D_2 = H_{2MAX} - H_2.$$

P_{ij} was the probability of an observation for a given cell in the matrix.

To determine if an observed D_2 value differed from random sampling, a Monte Carlo computer simulation was conducted on each matrix (Losey 1978). For each simulation the probability of obtaining a D_2 value at least as large as the one observed was calculated by randomly filling a matrix with the number of events observed. Since the observed frequency of behavioral acts in each row was highly skewed, during each simulation the sum of the number of events in each row was limited to the observed sums.

Spearman rank correlation (r) was used to compare the behavior of every individual in a colony to every other individual in that colony and to compare the behavior performed in each age category to every other age category. This provided an index of similarity (1 - r) between the behavior of pairs of individual workers and between different age categories that could then be used to perform a hierarchical cluster analysis using the complete linkage clustering method (Romesburg 1984). This resulted in clusters of workers related by a correlation value (r) at least as large as the value of the linkage line between the clusters. Principle component analysis (Frey and Pimentel 1978) was performed to determine the behavioral acts that were important in explaining the variation in behavior between workers.

RESULTS

Description of Behavioral Acts

A more detailed description of each act may be found in the references cited in the descriptions below.

Adult-Adult Interactions
(1) Antennate Nestmate--use the antennae to palpate the head, thorax, gaster, or wings of another female.
(2) Mouth Nestmate--use the mouthparts to lick or bite another female on the head, thorax, abdomen, or wings (Owen 1962; West Eberhard 1969).
(3) Dart at Nestmate--move rapidly or lunge, frequently while flipping the wings, toward another adult (West Eberhard 1969).
(4) Submissive--akinesis with the antennae, head, and body held close to the nest surface (Pardi 1948; West Eberhard 1969).
(5) Grapple--clash between two adults involving biting and attempted stinging (West Eberhard 1969).
(6) Trophallaxis--oral transfer of regurgitated liquid from one adult to another. The soliciting wasp frequently antennates the head of the other wasp during the transfer of liquid (West Eberhard 1969). Antennation occurring during trophallaxis is not counted as a separate act (that is, antennate nestmate [#1]).

Pulp Handling
(7) Land with Pulp--return to the nest with a mass of nesting material held in the mandibles.
(8) Malaxate Pulp--use the forelegs to hold and manipulate a mass of pulp while chewing it with the mandibles.
(9) Give Pulp--transfer all or part of a mass of pulp from the mandibles of one female to the mandibles of one to two other females simultaneously.

(10) Receive Pulp--acquire all or part of a
     malaxated mass of pulp from the mandibles
     of another female. This is the
     reciprocal of #9.
(11) Search--quickly walk around the nest with
     pulp held in the mandibles, antennating
     cells and the surface of the nest (West
     Eberhard 1969). This is usually followed
     by the addition of pulp to the nest (see
     #12-13 below).   .
(12) Initiate Cell--form a new nest cell by adding
     pulp to the nest. Pulp is worked with
     the mandibles and smeared out into a thin
     sheet on the nest (West Eberhard 1969).
(13) Add Pulp Cell--lengthen an existing cell by
     the addition of pulp. Pulp is chewed and
     worked with the mandibles onto the margin
     of a cell, while the antennae beat
     rapidly against the sides of adjacent
     cell walls (Owen 1962; West Eberhard
     1969).
(14) Add Pulp Petiole--enlarge the nest petiole by
     adding pulp. Pulp is worked with the
     mandibles and smeared out into a thin
     sheet on the sides of the petiole (Owen
     1962). During pulp addition the antennae
     beat continually against the nest petiole
     (West Eberhard 1969).

Prey Handling
(15) Land with Prey--return to the nest with a
     mass of prey held in the mandibles.
(16) Malaxate Prey--use the forelegs to hold and
     manipulate a mass of prey while chewing
     it. During malaxation the antennae
     quiver while they are held straight out
     at about a 40° angle to the front surface
     of the face.
(17) Give Prey--transfer all or part of a mass of
     prey to one or two other females
     simultaneously. The wasps frequently
     antennate each other during the transfer,
     but they often do not come into physical
     contact (Morimoto 1960; West Eberhard
     1969).
(18) Receive Prey--acquire all or part of a
     malaxated mass of prey from another

female (Owen 1962). During the transfer
of prey each female malaxates the mass of
prey, slowly dividing it into two or more
pieces. The antennae occasionally beat
against the mass of prey, but usually are
held stationary. This is the reciprocal
of #17.

(19) Feed Larva Prey--the adult, holding malaxated
prey in the mandibles, inserts her head
into·cells containing larvae and each
larva takes a bit of the solid malaxated
prey from the female (Owen 1962).
Successive acts of feeding a larva are
recorded as a single act.

Miscellaneous Adult Behavior

(20) Cell Inspection--walk over the nest surface
inserting the head into cells (Owen
1962). This behavior pattern includes
trophallaxis with the larva, antennation
of cell contents, and regurgitation of
liquids, acts that usually are not
distinguishable because they occur out of
view inside a cell. We do distinguish
between cell inspection, as defined
above, and feed larva prey, chew cell,
and brood abortion (see #19, #21, and
#22), acts that are easily observed.

(21) Abort Brood--insert the head into a cell
containing a larva, remove the larva from
the cell, and fly off the nest with it.

(22) Chew Cell--remove pulp or larval silk from a
cell either by using the mandibles to
bite at the margin of a cell or the cell
cap, or by inserting the head into a cell
and using the mandibles to chew away
larval silk or meconia.

(23) Lick Nest--move the mouthparts up and down
over the nest comb or petiole (Owen
1962). The tips of the antennae are
curved back with their dorsal surfaces in
contact with the nest and they are gently
rubbed over the nest surface.

(24) Rub Petiole--press the ventral surface of the
gaster against the nest petiole and
vigorously move it up and down (Post and
Jeanne 1981).

(25) Rub Nest--brush the ventral surface of the gaster lightly forward and backward over the surface of the comb (excluding the petiole) while walking over the nest surface.

(26) Abdomen Wag--move the gaster from side to side (usually 2-3 oscillations) over the surface of the nest while walking over the nest (Gamboa and Dew 1981). The side to side shaking of the gaster differentiates this act from rub nest (#25).

(27) Lateral Vibration--move the gaster rapidly and violently from side to side against the nest surface while the female is stationary (Gamboa and Dew 1981). This produces a loud rattling sound, and often involves shaking the whole body from side to side. The violent movement of the whole body and the production of sound differentiates this act from abdomen wag (#26).

(28) Antennal Drum--rapidly beat the antennae against the rim of a cell, producing a trilling sound (Pratte and Jeanne 1984). The whole body usually vibrates forward and backward.

(29) Fan--beat the wings steadily while stationary on the nest (Owen 1962). Fanning females frequently groom; this is not distinguished as an act separate from fanning and thus is not counted as a groom (see #37).

(30) Dart--run rapidly or lunge in one direction while flipping the wings. This is differentiated from dart at nestmate (see #3) by the fact that the movement is not directed toward another adult.

(31) Alert--orient, with forelegs often held close to the thorax and waving, toward an insect near the nest. This is a frequent reaction to returning foragers.

(32) Parasite Alarm--move rapidly and jerkily over the nest while flipping the wings (West Eberhard 1969). This is a response to the presence of the parasite Pachysomoides fulvus (Cresson)

(Hymenoptera: Ichneumonidae) near the nest.

(33) Fly Off-Attack--depart from the nest and fly towards an insect, usually a returning nestmate, near the nest.

(34) Fly Out--fly from the nest (rarely more than 15 cm) and then return. This is differentiated from fly off-attack (see #33) by the fact that no insect is visible near the nest.

(35) Land with Unknown--return to the nest without prey or pulp. This includes returning to the nest with water, nectar, or nothing. We do not differentiate between these items since water and nectar are carried in the crop and frequently it is not possible to ascertain the substance being carried.

(36) Groom--clean the body with legs and mouthparts (Owen 1962).

(37) Inactive--immobile for more than 15 seconds.

Frequencies of Behavioral Acts

The frequency of behavioral acts for workers are reported in Table 10.2. The probability that a next observed act had already been observed (Good's estimate of sample coverage, Fagen 1978) was 0.99997, indicating that most types of acts were observed.

Individual Variation in Behavior

There was significant divergence from independence of the complete behavioral repertory (37 acts) and worker identity for all 54 workers (Table 10.3A). The high D_1 value (1.966) relative to H_{MAX} (5.209) for the rows indicated that there was not an equal probability of occurrence of the behavioral acts, as is also evident in Table 10.2. Moreover, the divergence from 0 of the D_1 value for the columns of the matrix showed that the numbers of events observed for each worker were not equal. Within each colony, except Colony 14, there was divergence from independence (D_2) between behavior and worker identity

(Table 10.3B). In all of the Monte Carlo simulations none of the observed D_2 values were within the distribution of the randomly generated values. Thus, there was dependence between the behavioral act and worker identity, providing evidence of inter-worker variability in behavior.

Pair-wise comparison of the complete behavioral repertory of each worker to every other worker within a colony revealed a significant correlation among all pairs of workers (Spearman rank correlation: $r > 0.33$; $N = 37$; $P < .05$) except for workers 1 and 12 from Colony 11 and workers 19 and 20 from Colony 13. However, a large number of the behavioral acts, such as cell inspection, groom, inactive, etc., were performed at relatively high frequencies by all workers. Since the frequencies of these acts did not appear to correlate with other acts or have an apparent relationship to colony labor, they might obscure a real variation among workers in work performed. Thus, we confined the following analyses to those types of acts involved in pulp and prey handling, acts that clearly have an important role in colony functioning and productivity.

Information theory analysis of independence between the 13 behavioral acts involved in pulp and prey handling (see Table 10.2 for a list of the acts) and worker identity gave results similar to those for the complete behavioral repertory (Table 10.4A). Also, the behavioral patterns of numerous pairs of workers within each colony were not highly correlated ($r < 0.56$; $N = 13$; $P > .05$). Hierarchical cluster analysis (using the Spearman rank correlation as a similarity index) of the workers in the two colonies (3 and 13) having a large number of observed events revealed the similarity of behavior between pairs of workers (Figures 10.1 and 10.2). Although the number of events observed was small, clusters of workers similar to those for Colonies 3 and 13 were evident for Colonies 11 and 15. The workers in Colony 3 separated into two groups, one involved frequently in pulp and prey handling on the nest (workers 3, 6, 8, and 9) and the other in prey foraging and giving prey to a nestmate (workers 1, 2, 4, 5, and 7) (Figures 10.1 and 10.3). Although most of the wasps performed all of the tasks,

Table 10.2
Behavioral catalog of acts for P. fuscatus workers

Behavioral Type	Relative Frequency of Acts[a]	Frequency of Acts/ Worker/Hours on Nest	Number of Acts
Adult-Adult Interactions			
1. Antennate Nestmate	0.0219	0.933	268
2. Mouth Nestmate	0.0068	0.289	83
3. Dart at Nestmate	0.0071	0.303	87
4. Submissive	0.0111	0.473	136
5. Grapple	0.0008	0.035	10
6. Trophallaxis	0.0611	2.607	749
Pulp Handling			
7. Land with Pulp	0.0078	0.334	96
8. Malaxate Pulp	0.0097	0.414	119
9. Give Pulp	0.0049	0.209	60
10. Receive Pulp	0.0042	0.177	51

11.	Search	0.0140	0.599	172
12.	Initiate Cell	0.0001	0.003	1
13.	Add Pulp Cell	0.0118	0.501	144
14.	Add Pulp Petiole	0.0013	0.056	16

Prey Handling

15.	Land with Prey	0.0122	0.522	150
16.	Malaxate Prey	0.0275	1.173	337
17.	Give Prey	0.0121	0.515	148
18.	Receive Prey	0.0109	0.466	134
19.	Feed Larva Prey	0.0579	2.468	709

Miscellaneous Adult Behavior

20.	Cell Inspection	0.3297	14.001	4023
21.	Abort Brood	0.0004	0.017	5
22.	Chew Cell	0.0002	0.010	3
23.	Lick Nest	0.0017	0.073	21
24.	Rub Petiole	0.0042	0.181	52

(continued)

Table 10.2 (concluded)

Miscellaneous Adult Behavior

25.	Rub Nest	0.0016	0.070	20
26.	Abdomen Wag	0.0001	0.003	1
27.	Lateral Vibration	0.0002	0.007	2
28.	Antennal Drum	0.0007	0.031	9
29.	Fan	0.0086	0.365	105
30.	Dart	0.0006	0.024	7
31.	Alert	0.0097	0.414	119
32.	Parasite Alarm	0.0006	0.024	7
33.	Fly Off-Attack	0.0002	0.010	3
34.	Fly Out	0.0007	0.028	8
35.	Land with Unknown	0.0204	0.870	250
36.	Groom	0.2089	8.910	2560
37.	Inactive	0.1297	5.534	1590

[a]The frequencies are based on 12,255 acts recorded during 537.5 female-hours of observation of 54 different workers. Workers spent 53.46 percent of their time off the nest, resulting in 287.3 female-hours of observation of workers on the nest.

those workers in the pulp handling group were
significantly different from those in the prey
handling group (r < 0.560; N = 14; $\underline{P}$ > .05).
Clusters of workers distinguished by their roles
in prey foraging vs. pulp and prey handling were
also evident in Colony 13 (Figures 10.2 and
10.4). Workers 3 and 5 performed both prey and
pulp handling behavior, while workers 1, 2, 4, and
18 landed with prey and gave prey to a nestmate
more frequently than the former group. In
addition, one late-emerging wasp (20) rarely
partook in either pulp handling or prey handling,
was not observed landing on the nest with pulp or
prey, and rarely left the nest (0.1 percent of 7
hours of observation). A second late-emerging
wasp (19) did not perform any pulp or prey
handling acts and, in fact, was never observed off
the nest.
     The first and second principle components
explained 58.9 percent of the variance between
workers in the performance of the 13 pulp and prey
handling acts (Table 10.5). Within the first two
components the five prey handling acts loaded
strongly in a positive direction.

Age Polyethism and Individual Variation in Behavior

     The behavior of all workers combined was
dependent on age (Tables 10.3C and 10.4B). A
comparison of the frequency of the 13 pulp and
prey handling behavioral acts observed at each 2-
day age category to every other 2-day age category
using Spearman rank correlation revealed that the
behavior of the workers went through a transition
from ages 7-8 to ages 13-14 (Figure 10.5). In
addition, the behavior of the workers changed
again at about ages 41-42 days, although this may
be due in part to the small number of observations
of workers over 40 days old (Figure 10.6). Young
workers went through a transition from remaining
inactively on the nest, receiving pulp, and
receiving prey to foraging for pulp and prey,
feeding larvae, and building (Figures 10.7-
10.9). This trend was reversed, however, for

Table 10.3
Information theory analysis of independence (D_2) between the complete 37 behavioral act repertory and worker identities or 2-day age categories

Colony	No. of Rows	Behavioral Types		No. of colonies	Workers or Age		Second Order Independence		No. of Events
		H_{MAX}	D_1		H_{MAX}	D_1	H_{2MAX}	D_2	
				A. All Workers					
All	37	5.209	1.966	54	5.755	0.447	8.551	0.218[a]	12255
				B. Workers by Colony					
3	37	5.209	1.976	9	3.170	0.103	6.301	0.127[a]	3730
7	37	5.209	2.325	6	2.584	0.042	5.428	0.166[a]	900
11	37	5.209	2.051	12	3.585	0.305	6.439	0.327[a]	1970
13	37	5.209	1.927	8	3.000	0.382	5.901	0.089[a]	3257
14	37	5.209	2.712	5	2.322	0.131	4.688	0.155	289
15	37	5.209	2.007	9	3.170	0.190	6.182	0.293[a]	1219
16	37	5.209	1.871	5	2.322	0.233	5.428	0.147[a]	890

301

C. All Workers By Age (2-day interval)

All	37	5.209	1.966	25	4.644	0.606	7.281	0.134[a]	12255

[a]Monte Carlo computer simulations did not result in a D_2 value greater than these observed values.

Table 10.4
Information theory analysis of independence (D_2) between the 13 types of behavioral acts involved in pulp and prey handling and worker identities or 2-day age categories

Colony	No. of Rows	Behavioral Types		No. of Colonies	Workers or Age		Second Order Independence		No. of Events
		H_{MAX}	D_1		H_{MAX}	D_1	H_{2MAX}	D_2	
				A. Workers by Colony					
3	13	3.700	0.733	9	3.170	0.110	5.991	0.204[a]	793
7	13	3.700	0.991	6	2.585	0.592	4.703	0.375	77
11	13	3.700	0.642	12	3.585	1.283	5.361	0.449[a]	210
13	13	3.700	0.556	8	3.000	0.392	5.752	0.160[a]	601
14	13	3.700	1.826	5	2.322	0.887	3.310	0.260	22
15	13	3.700	0.914	9	3.170	0.968	4.988	0.260[a]	255
16	13	3.700	0.309	5	2.322	0.281	5.433	0.317[a]	179

B. All Workers By Age (2-day interval)

All	13	3.700	0.798	25	4.644	0.437	7.109	0.137[a]	1992

[a]Monte Carlo computer simulations did not result in a D_2 value greater than that observed.

COLONY 3

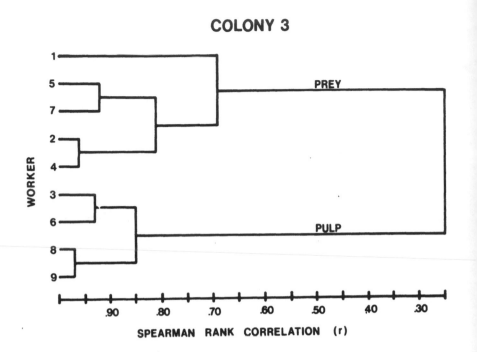

Figure 10.1. Complete linkage hierarchical cluster analysis of the workers in Colony 3. The clusters are based on the pair-wise comparison between all pairs of workers of the 13 behavioral acts associated with pulp and prey handling using the Spearman rank correlation value (r).

workers over 40 days old, which frequently were inactive and did not forage.

There was no detectable individual variation in age-related change in the behavior of workers other than the age at which the workers first left the nest. The first trip off the nest and the first observed landing on the nest with prey or pulp occurred at an older age for later emerging workers than for earlier emerging workers (first trip off the nest: $r = 0.59$; $\underline{P} = .0002$; $N = 51$ workers that were observed missing from the nest) (land on the nest with pulp or prey: $r = 0.56$; $\underline{P} = .0008$; $N = 34$ foragers). An analysis of Colonies 3 and 13 revealed that the age at which a worker first foraged for prey or pulp did not

COLONY 13

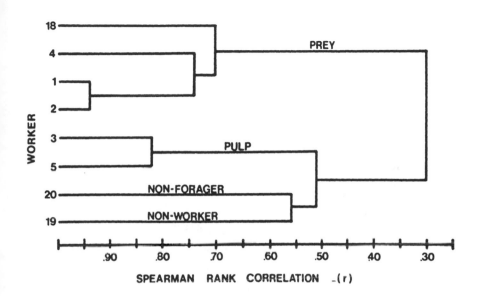

Figure 10.2. Complete linkage hierarchical cluster analysis of the workers in Colony 13. The clusters are calculated as in Figure 10.1

correlate with the order in which the wasps emerged (r = 0.17; P = .67; N = 9 for Colony 3 and r = 0.62; P = .17; N = 6 foragers on Colony 13). However, two marked wasps (19 and 20) from Colony 13 did not forage. If these two are included in the analysis as first foraging on the last day that they were observed, the age at the first foraging trip was correlated with order of emergence (r = 0.85; P = .008; N = 8).

Task Specialization and Individual Variation in Behavior

Workers of P. f. variatus frequently exhibited task partitioning in handling prey, but not in handling pulp. Returning prey foragers gave up the entire prey load to nestmates 44.2 percent of the time, while pulp foragers only gave

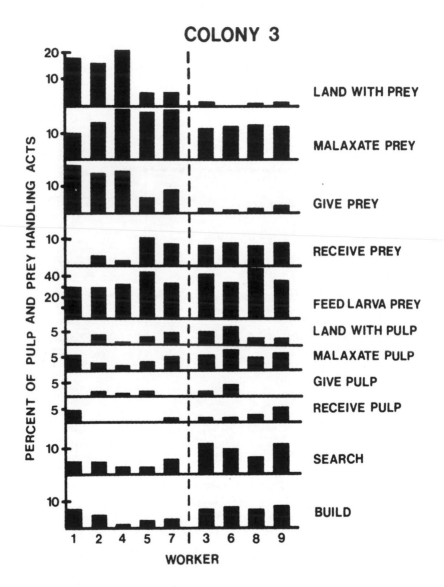

Figure 10.3. Relative distribution of the performance of the 13 pulp and prey handling acts for each worker in Colony 3. Workers classified as prey foragers and pulp and prey handlers are shown on the left and right side of the dotted line, respectively. Percent of each act is based only on the 13 pulp and prey handling acts.

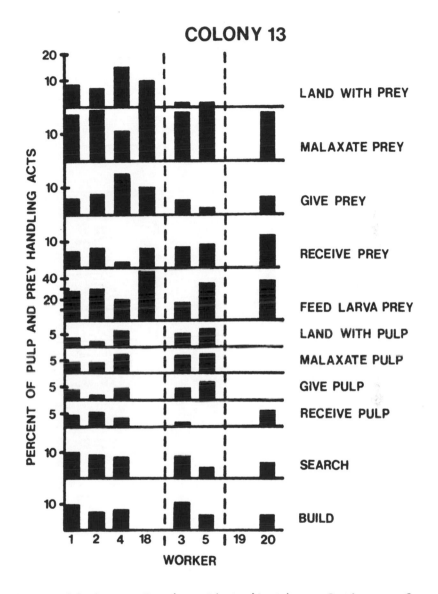

Figure 10.4. Relative distribution of the performance of the 13 pulp and prey handling acts for each worker in Colony 13. Workers shown on the left side of the dashed lines and between the two dashed lines are classified as prey foragers and pulp and prey handlers, respectively. Workers shown on the right side of the dashed lines are non-foragers. The percent of each act is based only on the 13 pulp and prey handling acts.

Table 10.5
Principle component analysis of the 13 behavioral
acts associated with pulp and prey handling
performed by 16 workers from Colonies 3 and 13[a]

Behavioral Type	Principle Component[b]			
	1	2	3	4
Land With Prey	0.71	-0.68	0.00	-0.01
Malaxate Prey	0.35	0.64	-0.27	-0.52
Give Prey	0.73	-0.60	0.09	-0.17
Receive Prey	-0.36	0.89	0.01	-0.08
Feed Larva Prey	0.30	0.70	-0.15	0.54
Land With Pulp	-0.76	-0.19	-0.57	-0.17
Malaxate Pulp	-0.87	-0.11	0.17	-0.07
Give Pulp	-0.55	-0.17	-0.65	-0.40
Receive Pulp	-0.25	-0.12	0.92	-0.11
Search	-0.85	-0.21	0.26	0.17
Initiate Cell	-0.29	0.08	-0.13	0.73
Add Pulp To Cell	-0.68	-0.60	0.03	0.04
Add Pulp To Pedicel	-0.21	0.52	0.61	-0.38
Eigenvalue	4.37	3.29	2.18	1.50
Percent Variance	33.6	25.3	16.8	11.5
Cumulative Percent Variance	33.6	58.9	75.7	87.2

[a]Worker 19 was not included since it did not
perform any of the 13 acts.

[b]Only the four principle components with
eigenvalues greater than one are shown.

up the whole load 6.4 percent of the time, usually
keeping at least part of the pulp load to build
(Table 10.6). Although workers as a whole tended
to partition prey handling tasks but not pulp
handling tasks, some workers might have exhibited
a greater or lesser tendency than other workers
for task partitioning. However, a comparison of
the frequency with which different workers gave up

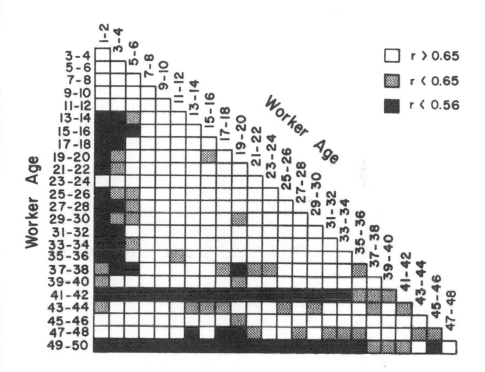

Figure 10.5 Comparison of the frequency of the 13 behavioral acts involved in pulp and prey handling of each 2-day age category to every other age category. The correlation coefficients of 0.56 and 0.65 correspond to P values of .05 and .02, respectively.

the whole load of prey or pulp to nestmates did not provide evidence of individual variation in task partitioning. For those workers for which there were more than 5 observations of prey handling there was no significant difference in the frequency with which the workers shared their loads, gave up all of their loads, or kept their loads (Colony 3: N = 3 workers; G = 1.19; d.f. = 4; P > .75; Colony 13: N = 2; G = 5.85; d.f. = 2; P > .05). The same was true for workers for which there were more than 5 observations of pulp handling (Colony 3: N = 3; G = 4.60; d.f. =

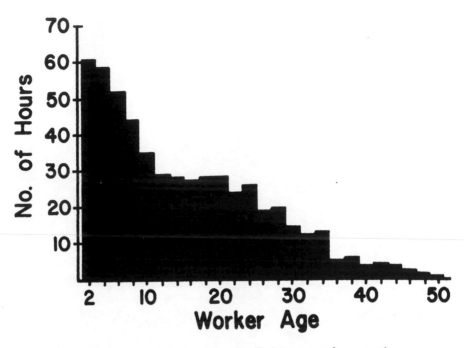

Figure 10.6 Total number of hours the workers
were observed at each age.

4; <u>P</u> > .25, Colony 13: N = 3; G = 6.62; d.f. = 4;
<u>P</u> > .10).

DISCUSSION

The results of this study show that there are
quantitative behavioral differences among workers
of <u>P. fuscatus variatus</u>. Information theory
analysis and Monte Carlo computer simulations
indicate that there is variation among workers
within a colony, not only in the relative
frequency of performance of the thirteen types of
prey and pulp handling acts, but also in the
frequency of performance of the types of acts in
the complete behavioral repertory. Within each
colony the workers specialize in certain acts.
Based on the thirteen acts involved in pulp and
prey handling, the workers can be divided into
three groups: prey foragers, pulp and prey

Figure 10.7 Average percent time the workers (N = 54 workers from 7 colonies) spent off the nest at each 2-day age interval.

handlers, and non-foragers. Variation in the performance of the prey handling acts, especially land with prey and give prey, appears to be responsible for explaining most of the variability between workers. However, all of the workers (except worker 19 in Colony 13) perform most or all of the different types of pulp and prey handling acts, and thus variation in behavior in P. f. variatus is quantitative, rather than qualitative. Although there is dependence between the acts within the prey handling and pulp handling categories (that is, the probability of performance of an act is dependent on the previous act), transitions between prey foraging and pulp and prey handling acts are independent (D.C. Post, unpublished data), indicating that the differences between pulp and prey handling workers are real and not due to dependence incurred from the method of observation. Future studies are necessary to

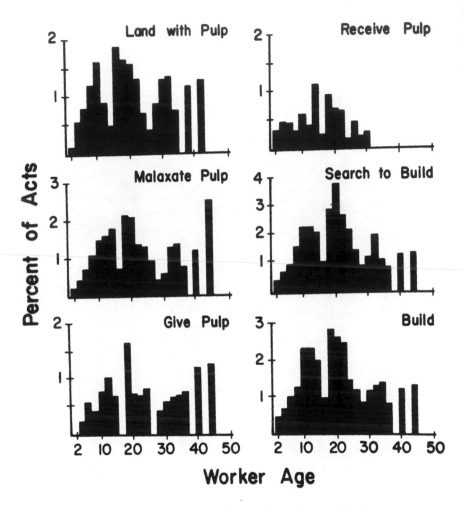

Figure 10.8. Relative distribution of the
    performance of 6 pulp handling acts at 2-day
    age intervals for all the workers (N=54)
    combined. Three acts--initiate cell, add
    pulp cell, and add pulp petiole--are combined
    in the distribution labelled build. The
    percent within each age category is based on
    all 37 types of acts.

determine the significance of this individual
variation among workers of Polistes.
    Owen (1962) and Strassmann et al. (1984)
describe similar patterns of variation among
workers of P. fuscatus fuscatus (F.) in Michigan

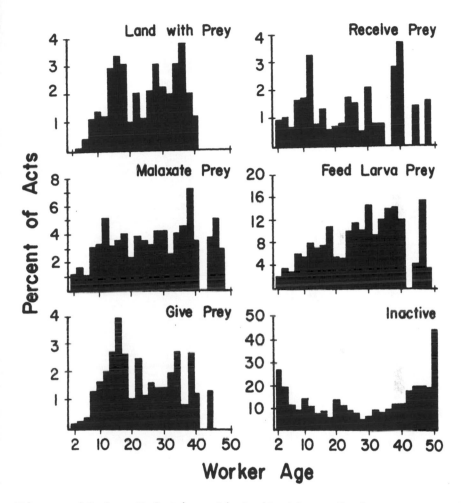

Figure 10.9. Relative distribution of the performance of the 5 prey handling acts and inactive at 2-day age intervals for all the workers (N=54) combined. The percent within each age category is based on all 37 types of acts.

and P. exclamans in Texas, respectively. From Owen's foraging data on P. f. fuscatus the females can be classified into four types: those that forage primarily for prey, those that forage primarily for pulp, those that forage for both prey and pulp about equally, and those that do not forage, or forage rarely. If data similar to that

Table 10.6
Load distribution behavior of returning prey and pulp foragers[a]

Initial Act	Kept Whole Load	Shared Load with Nestmate	Gave Whole Load to Nestmate
		Following Act	
Return to Nest With Pulp	49 (20.4%)	85 (35.4%)	106 (44.2%)
Return to Nest With Pulp	100 (49.3%)	90 (44.3%)	13 (6.4%)

[a]frequency that workers kept the whole load, shared the load with a nestmate, or gave up the whole load to a nestmate.

of the present study were available, foragers for both prey and pulp in Owen's study might be classified as either prey foragers or pulp handlers. In P. exclamans the greatest variability between workers is due to prey foraging; some females (foragers) are the primary prey foragers, although both foragers and non-foragers occasionally return to the nest with pulp (Strassmann et al. 1984).

Workers of P. f. variatus exhibit age polyethism that is based simply on a transition from on-nest activities or inactivity at young ages to foraging activities during middle age. Although old workers appear to decrease their foraging activity and revert to performing on-nest duties and remaining inactive, the number of observations of old wasps are too few to be certain that this transition is real and not a statistical aberration. As described for P. f. fuscatus (Owen 1962), P. exclamans (Strassmann et al. 1984), and P. metricus (Dew and Michener 1981), workers of P. f. variatus do vary in the age at which they first start to forage, with later emerging wasps beginning to forage at an older age than the first-emerging wasps. However, in P. metricus reared in the laboratory the first emerging worker is the primary forager while the other workers do little or no foraging unless the primary forager is removed (Dew and Michener 1981). We found that all females foraged in the relatively small colony (Colony 3), while on the larger colony (Colony 13) the last two wasps to emerge rarely foraged. Thus in our field colonies we did not have a predominant forager. Since prey foraging is much more time consuming and less successful in the field than in the laboratory, numerous workers are necessary to maintain the food and growth requirements of a colony, and unless a large number of workers emerge (as in Colony 13), all of the emerging workers forage.

There is little individual variation in the frequency with which the workers give up all of their foraged loads to nestmates. The frequency of task partitioning is probably based on colony size and the number of wasps on the nest (Jeanne 1986). Most Polistes colonies in northern areas normally have a small number of workers on the

nest at one time because they do not produce large
numbers of workers (range: 5-17; $\bar{x}$ = 9.1 workers
in this study) and because of forager mortality.
Thus, with few wasps on the nest there might be
little opportunity for task partitioning.

The wasps in this study, however, frequently
partition prey handling tasks but rarely partition
pulp handling tasks.   This may be because foraging
involves much more time and energy in locating a
prey item (caterpillar) than a pulp source (Post
1980), and the forager frequently cannot carry the
whole caterpillar back to the nest.   If the prey
item is too large to carry, a forager will divide
the prey and carry part of it back to the nest.
Considering the amount of time and energy invested
in finding one prey item and the chance that other
predators (especially ants) will find the
caterpillar (Post and Jeanne 1982), there is
probably considerable pressure for the forager to
retrieve the entire caterpillar.   Thus, there is
selection for the wasps to carry more than they
can handle back to the nest, quickly give the prey
to a nestmate and return to the site of the
remains of the caterpillar.   Pulp, on the other
hand, generally is found close to the nest and is
a stable, predictable resource.

Differences in pulp and prey availability may
also explain why there is more variability in the
performance of prey handling acts (specifically
land with prey and give prey to a nestmate) than
pulp handling acts.   Wasps that frequently forage
for prey presumably could become more efficient by
learning (1) the location of productive hunting
areas, (2) a search image for caterpillars, and
(3) how to efficiently subdue and handle prey.
Although some learning is important in pulp
handling, as stated above pulp is a stable,
predictable resource.

The adaptive significance of the non-working
wasps (workers?) is not clear.   It may be simply
that they are a reserve force and will forage if
colony growth or mortality of workers increases
the need for more foragers (Owen 1962; Dew and
Michener 1981).   Alternatively, it may be that
workers are executing personal options (Strassmann
et al. 1984).   Strassmann et al. (1984) argue that
the non-foragers in P. exclamans can gain personal

fitness by becoming egg-layers it the queen dies
or by forming a satellite nest. This may also be
the case in P. f. variatus, where a non-forager
may become an egg-layer if the original queen
dies. A third possibility is that in the present
study these females simply may be incipient queens
(gynes) that are produced near the end ot the
season. If these females do not forage, then they
may accumulate fat and overwinter. Thus they
could be a hedge against colony failure prior to
the burst of production of reproductives typically
seen in P. fuscatus.

We now have ample evidence that individuals
can be highly variable in their activity
patterns. Although it is important to remember
that in uncontrolled studies in natural field
conditions there are many variables (for example,
colony cycle, worker mortality, prey availability,
weather) that are likely to influence the behavior
of individual workers, it is nevertheless clear
that there are differences in behavior among
workers. Experimental studies designed to control
variables and/or to subject colonies to
contigencies are now needed to determine what
factors influence variability, how individuals
come to specialize on certain tasks, the
importance of colony needs and how individuals
determine colony need, and to determine the
importance of non-foragers (that is, are they
simply a reserve force or are these females acting
so as to maximize their individual reproductive
success?). With such studies of primitively
eusocial species we may gain insight into initial
stages in the evolution of the morphological caste
systems seen in some of the highly eusocial
insects.

SUMMARY

The behavior of 54 workers of Polistes
fuscatus variatus from 7 colonies located in the
field is quantified using the focal animal
sampling technique. A complete descriptive
catalog of 37 behavioral acts and their relative
frequency of occurrence is provided. An analysis
of differences in the lifetime behavioral

repertories among workers revealed that there are quantitative behavioral differences between workers within a colony and that the workers within each colony can be divided into three groups: prey foragers, pulp and prey handlers on the nest, and non-foragers. Workers exhibit age polyethism that is based primarily on a transition from on-nest activities or inactivity at ages less than 7-8 days to foraging activities at ages greater than 13-14 days. There was no detectable individual variation in age polyethism, other than the fact that the first trip off the nest occurred at an older age for workers emerging later in the colony cycle than for those emerging early in the cycle. The workers exhibited task partitioning while handling prey, but not in handling pulp; foragers returning to the nest with prey frequently gave the whole load of prey to a nestmate, while those returning with pulp usually held on to at least part of the pulp load and engaged in nest construction. There was, however, no detectable individual variation in task partitioning among workers.

ACKNOWLEDGMENTS

We thank Jeffrey R. Baylis for making available his computer programs and for his discussions on information theory. Research supported by the College of Agricultural and Life Sciences, University of Wisconsin, Madison, by Federal Hatch Project No. 2588 to Robert L. Jeanne, and by National Science Foundation grant BNS-8112744 to Robert L. Jeanne.

REFERENCES

Baylis, J.R. 1976. A quantitative study of long-term courtship. II. A comparative study of the dynamics of courtship in two new world cichlid fishes. Behaviour 49:227-267.
Dew, H.E. 1983. Division of labor and queen influence in laboratory colonies of Polistes

metricus (Hymenoptera: Vespidae).
Zeitschrift für Tierpsychologie 61:127-140.
Dew, H.E., and Michener, C.D. 1981. Division of
labor among workers of Polistes metricus
(Hymenoptera: Vespidae): laboratory
foraging activities. Insectes Sociaux 28:87-
101.
Fagan, R.M. 1978. Repertoire analysis. In
Quantitative ethology, ed. P.W. Colgan, 25-
42. New York: John Wiley and Sons.
Frey, D.F., and Pimentel, R.A. 1978. Principle
component analysis and factor analysis. In
Quantitative ethology, ed. P.W. Colgan, 219-
245. New York: John Wiley and Sons.
Gamboa, G.J., and Dew, H.E. 1981. Intracolonial
communication by body oscillations in the
paper wasp, Polistes metricus. Insectes
Sociaux 28:13-26.
Jeanne, R.L. 1980. Evolution of social behavior
in the Vespidae. Annual Review of Entomology
25:371-396.
Jeanne, R.L. 1986. The evolution of the
organization of work in social insects.
Monitore Zoologico Italiano (N.S.) 20:119-
133.
Kasuya, E. 1983. Behavioral ecology of Japanese
paper wasps, Polistes spp. IV. Comparison
of ethograms between queens and workers of P.
chinensis antennalis in the ergonomic
stage. Journal of Ethology 1:34-45.
Kolmes, S.A. 1985. An information-theory
analysis of task specialization among worker
honey bees performing hive duties. Animal
Behaviour 33:181-187.
Lehner, P.N. 1979. Handbook of ethological
methods. New York: Garland STPM Press.
Losey, G.S. 1978. Information theory and
communication. In Quantitative ethology, ed.
P. W. Colgan, 43-78. New York: John Wiley
and Sons.
Morimoto, R. 1960. On the social cooperation in
Polistes chinensis antennalis Perez.
(Studies on the social Hymenoptera of
Japan. IX). Kontyû, Tokyo 28:198-206.
Oster, G.F., and Wilson, E.O. 1978. Caste and
ecology in the social insects. Princeton,
New Jersey: Princeton University Press.

320

Owen, J. 1962. The behavior of a social wasp
Polistes fuscatus at the nest, with special
references to differences between
individuals. Ph.D. diss., University of
Michigan, Ann Arbor.

Pardi, L. 1948. Dominance order in Polistes
wasps. Physiological Zoology 21:1-13.

Pardi, L. 1951. Ricerche sui Polistini 12.
Studio della attivitá e della divisione di
lavoro in una societá di Polistes gallicus
(L.) dopo la comparsa delle operaie. Archivo
Zoologico Italiano 36:363-431.

Post, D.C. 1980. Chemical defense in the
temperate social wasp, Polistes fuscatus
(Hymenoptera: Vespidae). Masters thesis,
University of Wisconsin, Madison.

Post, D.C., and Jeanne, R.L. 1981. Colony
defense against ants by Polistes fuscatus
(Hymenoptera: Vespidae) in Wisconsin.
Journal of the Kansas Entomological Society
54:599-615.

Post, D.C., and Jeanne, R.L. 1982. Rate of
exploitation of arboreal baits by ants in an
old-field habitat in Wisconsin. American
Midland Naturalist 108:88-95.

Pratte, M., and Jeanne, R.L. 1984. Antennal
drumming behavior in Polistes wasps
(Hymenoptera: Vespidae). Zeitschrift für
Tierpsychologie 66:177-188.

Reeve, H.K., and Gamboa, G.J. 1983. Colony
activity integration in primitively eusocial
wasps: the role of the queen (Polistes
fuscatus, Hymenoptera: Vespidae).
Behavioral Ecology and Sociobiology 13:63-74.

Romesburg, H.C. 1984. Cluster analysis for
researchers. Belmont, California: Lifetime
Learning Publications.

Strassmann, J.E., Meyer, D.C., and Matlock, R.L.
1984. Behavioral castes in the social wasp,
Polistes exclamans (Hymenoptera:
Vespidae). Sociobiology 8:211-224.

West Eberhard, M.J. 1969. The social biology of
polistine wasps. Miscellaneous Publications,
Museum of Zoology, University of Michigan
140:1-101.

Yoshikawa, K. 1963. 1. Introductory studies on
the life economy of polistine wasps. II.

Superindividual stage. 2. Division of labor among workers. Japanese Journal of Ecology 13:53-57.

11

Age Polyethism and Individual Variation in *Polybia occidentalis,* an Advanced Eusocial Wasp

Robert L. Jeanne, Holly A. Downing, and David C. Post

The swarm-founding vespids, despite their comprising more than 24 genera and approximately one-third of the social species of wasps, have been little studied. This is due in part to their tropical distribution, away from the prying eyes of the majority of social insect biologists, and in part due to the difficulties of studying behavior of what may be thousands of individuals in a 3-dimensional enclosed nest. The phenomenon of worker polyethism is a case in point. Since Butler (1609) first described it in honey bees, polyethism in social insects has been the subject of uncounted studies (for reviews see Wilson 1971; Oster & Wilson 1978; Brian 1979). Worker polyethism in the swarm-founding wasps, however, has been touched on in only three recent studies, all of them unpublished dissertations. Naumann (1970) found evidence of temporal division of labor in Protopolybia acutiscutis (=pumila). Simoes (1977), in the most detailed study so far, used marked workers of known age to demonstrate temporal division of labor in Stelopolybia pallipes and S. exigua. Forsyth (1978) used workers of known relative age to demonstrate the same in Polybia occidentalis. All have shown that the typical pattern of age polyethism prevails in these species: young workers engage in tasks at the nest, then shift to foraging as they get older.

Each of these studies of division of labor relies on the averaged behavior of a sample of workers in each caste. This approach, which

enables the inclusion of data collected from large
numbers of individuals, has revealed much about
worker polyethism. Nonetheless, by lumping the
behavior of many workers in a caste, it has the
weakness of obscuring what might be significant
differences among the individuals making up each
caste.

To take an example, Seeley's (1982) excellent
analysis of age polyethism in honey bees reveals
that the 13 most common tasks fall into five
groups, each comprising tasks sharing a common age
range over which relative probability of task
performance is high. The first of these
("cleaning cells") ranges from age 0 to 25 days,
but has its highest concentration in the range of
0 to 5 days. The second set consists of four
tasks occurring at the nest center and peaking in
the age range of 2 to 11 days. Thus, these tasks
overlap broadly with cell cleaning. Seeley
defines age caste 1 as specialists on cell
cleaning and with an age range of 0 to 2 days.
Age caste 2 has an age range of 2 to 11 days and
performs both task sets (as well as a third).
Because Seeley treated his marked workers as a
population, his results may obscure what, on the
individual level, is actually a discrete division
of these two task-sets among the two age castes.
That is, the data on cell cleaning performed at
older ages may have been contributed by workers
that switched to the second task-set later than
the average worker.

We undertook our study of age polyethism in
Polybia occidentalis because we were interested in
knowing more detail about the role of age in
determining worker behavior and because we wanted
to begin to investigate the degree of individual
variability among members of a caste with respect
to their role performances. We confirm that age
polyethism does occur and is quite pronounced,
that the shift from one temporal caste to another
is more abrupt than is revealed by studying groups
of workers rather than individuals, and that there
is a tremendous amount of individual variability
with respect to the details of age-related
specialization.

METHODS

The work was conducted in the field at
Hacienda La Pacifica, near Cañas, Guanacaste
Province, Costa Rica. We located colonies of P.
occidentalis in shrubbery in pastures and other
areas of low second growth vegetation and observed
them in situ. The results reported here come from
two colonies (110 and 125) studied during June and
July 1984.

To obtain workers of known age we collected
colonies, separated the combs, and incubated them
in plastic bags at ambient temperature. Each
morning we removed the adults that had eclosed
from these combs over the previous 24 hours.
These were chilled, then marked on the thorax with
colored typing correction fluid to indicate the
age cohort to which they belonged. Dots of model
airplane enamel applied to this background color-
coded individual identity. Length of the right
forewing was measured to the nearest 0.1 mm with a
vernier caliper. Within 1 h after marking, the
wasps were rechilled and introduced into the
observation nest by pushing them through the nest
entrance. We introduced a cohort of 14-29 of
these individuals into each observation nest every
other day. Colony 110 received 6 cohorts
beginning on June 11 and Colony 125 received 5
cohorts beginning on June 12. Upon introduction,
the marked wasps were antennated and groomed by
resident wasps, but were then accepted (with rare
exceptions) into the observation nests, where they
went on to become functional workers. Of 138
introductions into Colony 110 (103 in Colony 125),
111 (88 in 125) were "successful," that is the
wasps survived long enough to yield entries in our
records of behavior. The unsuccessful
introductions may have (1) died as a result of
injury sustained in marking or of paint in the
joints, (2) been driven out by the resident wasps,
or (3) left the nest of their own accord shortly
after introduction.

Beginning on the day of introduction of the
first cohort, we sampled the behavior of the
marked individuals at the nest in two ways.
First, we scan sampled their behavior at 5-min
intervals. During each scan the behavior of every

marked individual visible on the nest was recorded; if an individual could not be found it received the entry "not visible." Frequency of acts is reported as the number of acts per 1000 scans, where "scan" in this context is the entry for a single worker at each 5-min interval. Second, departures from and arrivals at the nest by marked individuals were recorded as they occurred. Solid loads brought in by returning foragers were directly identifiable as pulp or prey. Liquid crop-loads of water were distinguished from nectar by differences in trophallaxis behavior at the nest (Hunt et al. 1987).

The observation day was broken up into four 3-h periods, running from 0530 h to 1730 h. Each colony was observed for two 3-h periods per day according to the following 2-day recycling schedule: Day 1: 0530-0830 h, 1130-1430 h; day 2: 0830-1130 h, 1430-1730 h. This ensured that the full 12-hour daylight period was covered every two days for each colony. Observations were continued at this level through 5 July on Colony 110 and 8 July on Colony 125. By these dates survivorship of marked workers was low, so observations were reduced from 6 to 4 h/day on each colony and continued until 29 July (125) and 30 July (110). The two colonies were observed for a total of 345 hours. One of us (HAD) made all the observations on Colony 110. Colony 125 was observed by RLJ through July 11 and by DCP thereafter.

Colonies were left unmanipulated except that every two days we removed the lower envelope from each nest to elicit nest repair. In this way we provided a more or less continuous low level demand for several important and observable behavioral acts against which we could measure the effects of worker age. It also enabled us to obtain small samples of observations of activities inside the nest.

Results were analyzed as matrices of behavior of each individual vs. age. Data were lumped into 2-day age intervals to smooth differences in behavior due to the time of day during our 2-day observation schedule.

BEHAVIORAL ACTS

We recorded 67 behavioral categories. Of
these, 31 were used in the analysis reported
here. The remaining 36 included those that were
not social acts, whose functions were ambiguous or
unknown, or did not occur frequently enough to
make analysis useful for the present purpose.
Since the 31 we used are social acts, i.e.
constitute work contributed to the welfare of the
colony, we also refer interchangeably to these
acts as "tasks," in accordance with current usage
(e.g., Wilson 1980; Seeley 1982). Seeley used the
term "task-set" to refer to the group of tasks
whose performances are spatially juxtaposed in the
nest. We use the term to refer to tasks that have
a more functional relatedness (although the task-
set "foraging" is spatially distinct from the
others in taking place off the nest). Finally,
"role" is used herein to mean the one or more
task-sets performed by a given age caste.
The 31 acts/tasks used in our analysis are
divided into 6 task-sets and defined below.

Nest Building

1. Walk with pulp--move around the nest with
    pulp held in the mandibles.
2. Build cell--use the mandibles to work a
    load of nest material (pulp) into the wall
    of a brood cell, increasing its length.
3. Build envelope--use mandibles to work a
    load of pulp onto the edge of the nest
    envelope, either in nest enlargement or
    repair.
4. Build surface--use mandibles to tamp a load
    of pulp onto the outer surface of the
    completed envelope.
5. Malaxate pulp--use the forelegs to hold and
    manipulate a mass of pulp while masticating
    it with the mandibles.
6. Receive pulp--acquire all or part of a pulp
    load from another worker.
7. Suck wet pulp--imbibe moisture from still-
    wet pulp recently added to the nest.
8. Fan wet pulp--stand over recently added

pulp while buzzing wings continuously.

Nest Maintenance Inside Nest

9   Clean cell--remove exuviae/silk cap from recently evacuated brood cell.

10.   Carry trash--move over nest carrying solid waste in the mandibles.

11.   Drop water from nest--extend anterior part of body out from lower part of nest, regurgitate water from crop, and let droplet fall to the ground.

Nest Maintenance Outside Nest

12.   Fan nest--stand in place on nest envelope while vibrating wings steadily as if flying.   Done in response to elevated nest temperature.

13.   Lick envelope--repeatedly extend maxillae and labium and pass them over surface of envelope.

14.   Suck water from nest--imbibe water from nest surface wetted by rain.

Brood Care

15.   Receive prey--acquire all or part of prey load from a nestmate.

16.   Malaxate prey--use the forelegs to hold and manipulate a mass of prey while masticating it.

17.   Walk with prey--move over the nest carrying a prey load in mandibles.

18.   Feed larva prey--place bit of prey on mouthparts of larva.

19.   Feed larva liquid--regurgitate crop liquid and present it to larva.

20.   Inspect egg/larva--put head into open brood cell.

Defense

21.   Alarm--spread wings suddenly while raising anterior of body.

22.   Attack--fly from nest at intruder.

23. Chew/lick twig--use mouthparts to manipulate surface of twig supporting nest in response to presence of ants on twig.
24. Fan wings at ant on nest--buzz wings in short bursts in response to ant on nest surface.
25. Fan wings at flying insect--buzz wings in short bursts in response to insect flying close to nest.
26. Flick wings at flying insect--flip wings repeatedly at small insect flying close to nest.
27. Remove ant--use mandibles to remove ant from nest surface.

Foraging

28. Land with prey--return to nest with solid food in mandibles.
29. Land with pulp--return to nest with pellet of nest material in mandibles.
30. Land with water--return to nest with water in crop.
31. Land with nectar--return to nest with nectar or honeydew in crop.

COLONY BEHAVIOR DURING OBSERVATION PERIOD

By the time we began observations in early June, both colonies had resumed rearing brood following a period of no brood rearing during the dry season just ended. However, each colony produced only one generation of offspring. As the last of these offspring eclosed, each colony evacuated its nest, emigrated as a swarm to a new site nearby, and constructed a new nest in which it immediately began rearing a new cohort of brood. Colony 110 emigrated on 9 July and Colony 125 on 15 July. This abandoning of the old nest and construction of a new in July was apparently part of the normal seasonal cycle, for 10 of 12 other colonies we had under observation did the same.
Dramatic changes in the behavior of each colony during the course of the observation period were correlated with this phenomenon. At the

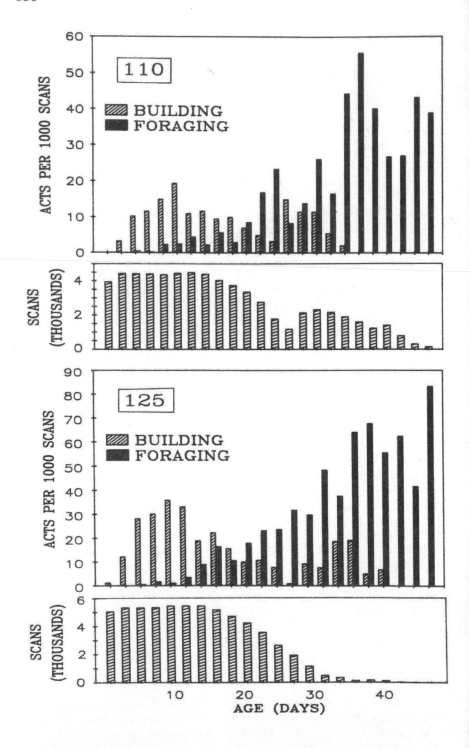

beginning of the period workers in each colony
foraged for prey to feed their large population ot
larvae and quickly repaired holes in the envelope
made by us. As the brood pupated, however, prey
foraging dropped off and holes in the nest were
repaired more and more slowly. As the last of
these offspring eclosed, the colony began leaving
the experimentally imposed nest damage
unrepaired. Several days later the adult
population clustered on the outside of the
envelope, and within a day or two left the nest.
At the new nest, begun immediately, nest
construction occupied a large proportion of the
workers for several days until several combs had
been built and covered with envelope. Then the
workers turned increasingly toward prey foraging
to feed the larvae that had begun to hatch. At
the end of observations on both nests, all the
behavior patterns associated with brood rearing,
nest maintenance, and defense were in evidence.

TEMPORAL SEPARATION OF ON-NEST AND OFF-NEST ROLES

For this analysis we used only workers that
were observed to perform at least two foraging
trips. The criterion of two trips was chosen
arbitrarily to qualify a worker as a forager. We
used nest building as an indicator of the on-nest
role because it was represented by large sample
sizes. Plotting the frequency distribution of
building and foraging tasks against age clearly
reveals age polyethism with respect to these two
task-sets (Figure 11.1). There is a concentration
of building tasks during younger ages, and a
concentration of foraging at older ages, but with

Figure 11.1. Frequencies of building and foraging
    behavior as a function of worker age.
    Combined sample of all workers performing at
    least 2 foraging trips. Lower graph gives
    the number of scan samples on which the upper
    graph is based. Colony 110: N = 31
    workers. Colony 125: N = 38 workers.

considerable overlap of the two task-sets in the age range of 6-38 days.

This overlap has two components. First is intra-individual overlap. That is, each individual may turn gradually from building to foraging, passing through several days when it performs both task-sets before switching exclusively to foraging. Figure 11.2 shows the distribution among workers of days of overlap of building and foraging. The second component is inter-individual overlap. That is, even if individuals do not overlap performance of the two task-sets, but make an abrupt and complete switch from building to foraging, the age at which the switch occurs may vary among individuals. Either component could act alone to yield the overlap at the group level seen in Figure 11.1.

In order to estimate the relative importance of these two sources of overlap, we replotted the same data in a different way. For each worker we estimated the age at which it switched from building to foraging. Some workers made the switch abruptly, such that foraging was not begun prior to the end of performance of on-nest tasks (within the limits of resolution of our 2-day age intervals). Most, however, overlapped the two roles by some number of days. For these, we defined the switch age as the 2-day age interval nearest the point at which the number of building acts recorded subsequent to it equalled the number of foraging acts recored prior to it. Figure 11.3 shows the distribution of switch ages among workers in each colony.

We then plotted the task distribution for all individuals in our sample so as to center their individual ages of switching on the same point on the x-axis (Figure 11.4). Plotting the data in this way illustrates the relative abruptness of the switch from building to foraging by individual workers. Other categories of on-nest task-sets terminate at about the same time as building. For each task-set there is at least a two-fold change in frequency across the switch point.

Plotting the task frequency distribution in this way removes the overlap in building and foraging due to the second component, namely, individual variance in age of switching, leaving only intra-individual overlap. The resulting task distribution is significantly different from that in Figure 11.1. (Colony 110: x^2 = 146.2, df = 3, P << .001. Colony 125: x^2 = 229.7, df = 3, P << .001.)

We next estimated the relative contributions of the two components by summing the building and foraging acts contributing to overlap within individuals (overlap depicted in Figure 11.2) and dividing by the sum of building and foraging acts contributing to overlap from both sources (overlap depicted in Figure 11.1). The resulting values were 0.36 and 0.13 for Colonies 110 and 125, respectively. Thus we estimate that 36 and 13 percent, respectively, of the overlap at the group level in the two colonies is due to the overlapping of building and foraging acts within the ontogeny of individual workers, the rest being due to differences in the ages at which individuals made the switch.

Workers foraged for four materials: nectar, pulp, prey, and water. The distribution among foragers of trips for these materials varied considerably (Table 11.1). Is there age polyethism with respect to foraging specialty? Sample sizes of water and prey were too small to analyze, but there was a pattern with respect to nectar and pulp. Of the 6 workers (both colonies) that made at least one trip for each material, 5 foraged for pulp first, shifting to nectar after several days (Figure 11.5). The frequency distributions for pulp and nectar for all 6 workers combined are significantly different from one another (x^2 = 106, df = 4, P < .005). Two other workers that started out as pulp foragers disappeared 4 and 8 days into their foraging periods, apparently before they could switch to nectar. The mean time of switching from pulp to nectar was 7.6 days after switching to foraging (range = 2-10 days).

Each of the 5 workers that foraged first for pulp started their foraging careers either while the old nest was still being repaired or after they arrived at the new nest site, i.e. while nest building was a colony activity. This is not surprising, of course, but it should not be taken to mean that the pulp-to-nectar sequence is

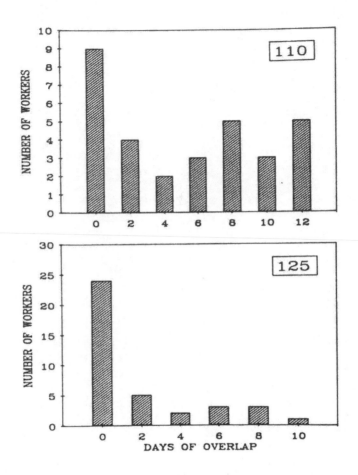

Figure 11.2.   Frequency distribution of the number
    of days during which both building and
    foraging were performed by workers.   "0"
    means that there was no overlap of building
    and foraging within the 2-day age intervals
    used.   Colony 110:   Mean = 5.29 days,
    SD = ±4.58 days, N = 31 workers.
    [Eliminating workers having zero overlap:
    Mean = 7.45 days, SD = ±3.61 days, N = 22
    workers.]   Colony 125:   Mean = 1.84 days,
    SD = ±2.95 days, N = 38 workers.
    [Eliminating workers having zero overlap:
    Mean = 5.00 days, SD = ±2.80 days, N = 14
    workers.]

entirely an epiphenomenon of the shape of the
labor needs in the colony at the time these
workers happened to begin foraging. Many other
workers in both colonies started their foraging
lives as nectar foragers, yet failed to switch
later to pulp foraging, even during construction
of the new nest, when building and pulp foraging
were major activities occupying a large proportion
of the worker force (Figure 11.6). Thus the

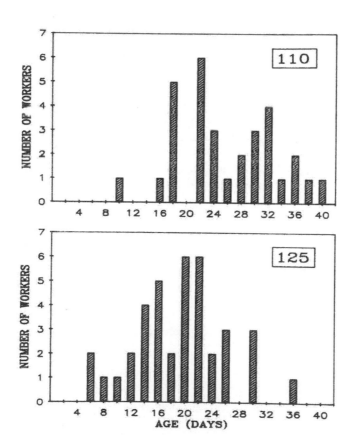

Figure 11.3.   Frequency distribution of age at
    which workers switched from building tasks to
    foraging. Colony 110: Mean (±SD) = 25.7
    (±7.5) days. N = 31 workers. Colony 125:
    Mean (±SD) = 19.6 (±6.4) days.

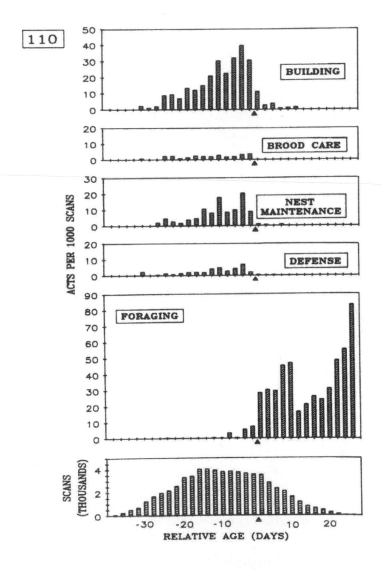

Figure 11.4. Frequency distribution of on-nest tasks and foraging. The x-axis represents the days of age before and after the switch (arrow) from the one role to the other for

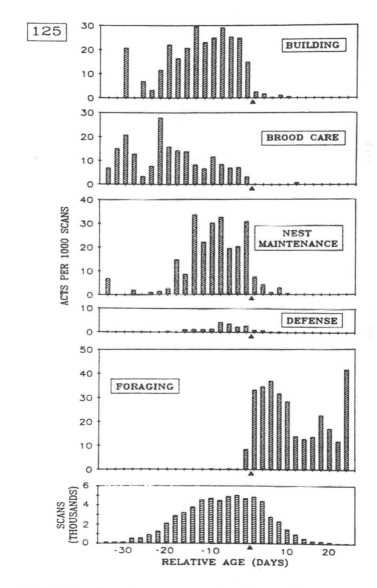

each individual. Lower graph gives the number of scan samples on which the data in the upper graph are based. Colony 110: $\underline{N}$ = 31 workers. Colony 125: $\underline{N}$ = 38 workers.

338

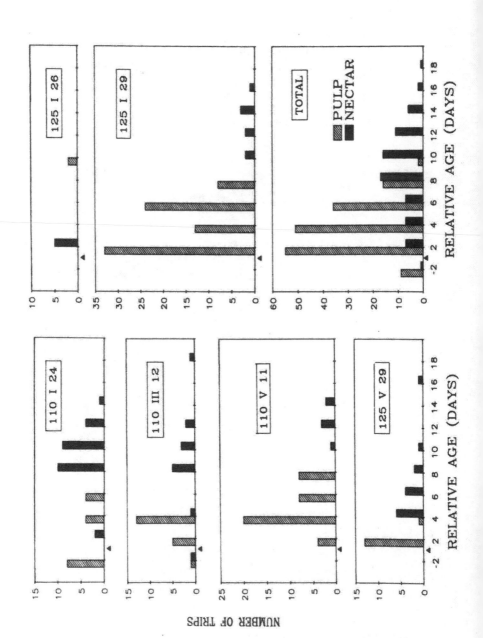

Figure 11.5. Frequencies of pulp and nectar foraging for the six workers (both colonies) that performed both tasks. The x-axis represents the days of age before and after the switch (arrow) from on-nest tasks to foraging for each individual. Bottom graph gives the sum of all 6.

Table 11.1
Distribution among workers of foraging trips for each type of material

Colony	Foraged Material	Number of Workers	Total Trips	Number of Trips/Worker		
				Mean	SD	Range
110	Nectar	31	381	12.3	± 9.3	1-39
	Pulp	6	80	13.3	±14.4	1-41
	Prey	3	3	1.0	± 0.0	1
	Water	14	118	8.4	±19.1	1-77
125	Nectar	36	449	12.5	± 9.7	1-39
	Pulp	7	164	23.4	±25.6	1-78
	Prey	1	1	1.0	± 0.0	1
	Water	8	8	1.0	± 0.0	1

appearance of pulp foraging first, followed by
nectar foraging, occurred in several workers, but
nectar followed by pulp did not (the one possible
exception was I 26 of Colony 125, but she made
relatively few trips of either kind [Figure
11.5]), despite the opportunity for both within
the context of labor needs in both colonies during
the foraging careers of the majority of the
workers. This suggests that there is an intrinsic
tendency to forage for pulp first, but that it
manifests itself only in some individuals.

Worker wing length did not predict the age of
switching from nest duties to foraging (Table
11.2). Similarly, there was no difference in wing
length between foragers (those observed to make 2
or more trips in their lifetimes) and non-foragers
(those living at least 25 days but not observed to
make any foraging trips) (t-test; Table 11.3).

The event of colony emigration may have had
some effect on when individuals switched from on-
nest to off-nest roles. Although workers switched
before, during, and after preparations for
emigration began and even after emigration was
completed (Figure 11.6), the curves of cumulative
numbers switching to foraging plotted against date
show an acceleration in numbers switching as the
preparation to emigration phase was entered in
each colony, especially in Colony 125. Yet in
Colony 110 only 17 of 31 (55%) had made the switch
by emigration, while in Colony 125 36 of 38 (95%)
did.

The foregoing may leave the impression that
workers differed from one another mainly in the
rates at which they progressed through first on-
nest then off-nest activities. This was not the
case. Of the 56 workers of Colony 110 that lived
14 days or more, only 31 (55 percent) exhibited
the "normal" sequence of in-nest, on-nest, and
off-nest phases. On Colony 125 the number was 36
of 52 (69 percent). Two more began foraging
without going through an on-nest period. The
remaining workers in each colony did no foraging
(made one trip or no trips), that is went through
only the on-nest phase. The possibility that non-
foragers were merely those that did not survive
long enough to switch is not supported: In Colony
125 the mean longevities of foragers and non-

Table 11.2
Regression of age of switching from on-nest tasks to foraging on worker body size (wing length)

Colony	N	Age (Days)			Wing Length (mm)			R	t	p
		Mean	SD	Range	Mean	SD	Range			
110	31	25.7	±7.5	10-40	7.5	±0.2	7.1-7.9	0.20	0.25	.80
125	38	19.6	±6.4	8-36	7.5	±0.2	7.2-8.1	0.20	-1.20	.24

Table 11.3
Comparison of foragers and non-foragers with respect to body size
(wing length).

| Colony | Specialty | N | Wing Length (mm) | | | t | P |
			Mean	SD	Range		
110	Forager	31	7.5	±0.2	7.1-7.9		
	Non-forager	6	7.5	±0.1	7.4-7.8	-0.108	.91
125	Forager	38	7.5	±0.2	7.2-8.1		
	Non-forager	6	7.4	±0.3	7.1-7.9	1.05	.30

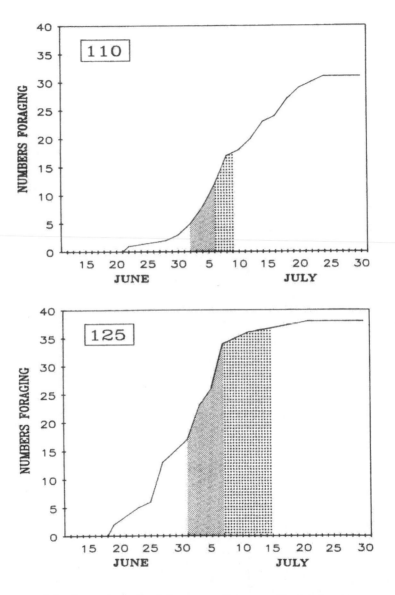

Figure 11.6. Cumulative numbers of workers
switching from on-nest tasks to foraging.
Fine shading = period of preparation for
emigration: nest damage repaired slowly.
Coarse shading = period of emigration: nest
damage not repaired. Colony 110: $\underline{N}$ = 31
workers. Colony 125: $\underline{N}$ = 38 workers.

foragers did not differ significantly [mean
(±SD): foragers = 28.4 (±7.8) days; non-foragers
= 29.2 (±10.6) days. $\underline{P}$ = .798, $\underline{df}$ = 50], although
they did in Colony 110 [foragers = 35.0 (±10.6)
days; non-foragers = 25.2 (±13.4) days. $\underline{P}$ = .004,
$\underline{df}$ = 55].

Not only did the time spent in each phase
vary among workers (see above results, Figure
11.3), but levels of activity varied widely as
well. A few workers rarely emerged from the nest
during their entire lives, yet were collected with
the colony at the end of the study. Details of
individual differences in activity level will be
treated elsewhere.

EVIDENCE FOR A TEMPORAL SEPARATION OF IN-NEST AND
ON-NEST ROLES

Although we were not able to observe the
behavior of workers inside the nest with any
thoroughness, the youngest workers may represent a
temporal caste that is distinct from the on-nest
caste characterized above. This is suggested by
the fact that the youngest workers rarely were
seen outside the nest. At several days of age the
typical worker became much more frequently
recorded in our scans. In many cases this
increase in "visibility" was abrupt.

We investigated this by analyzing the
correlation between appearance outside the nest
and onset of on-nest tasks. Again, the nest
building task-set was selected to define on-nest
activity because its component acts are performed
on the outside of the nest and because they are
represented by large sample sizes. Because the
age at which building was first observed varied
from individual to individual, we plotted the data
with the age of onset of building centered on the
same point on the $\underline{x}$-axis. Our samples consisted
of all workers that survived 14 days or more. As
can be seen in Figure 11.7, building activity
started abruptly in workers of both colonies.

The other task-sets, plotted so as to center
them on the same switch age, have roughly the same
distribution as does building. There is little
evidence of any temporal separation among them.

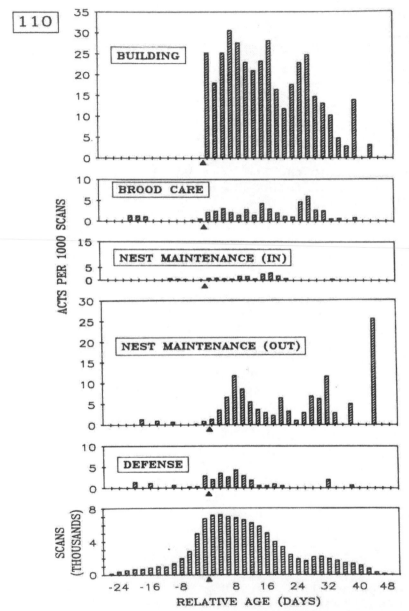

Figure 11.7. Frequency distribution of on-nest
    task-sets. The x-axis represents the days of
    age before and after the switching age
    (arrow) for each individual. See
    "Behavioral Acts" for the list of acts in
    each task-set. Bottom graph gives the number

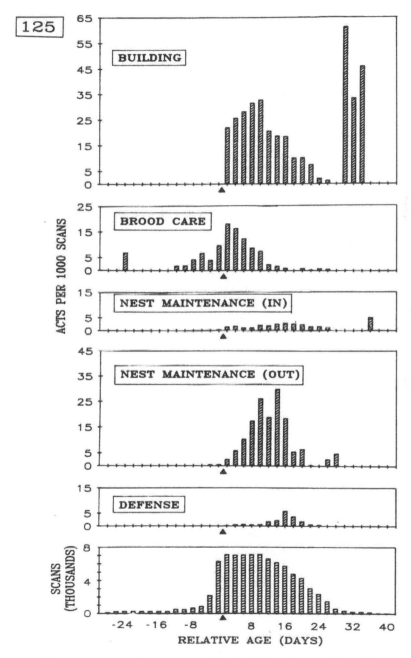

of scan samples on which the data in the
upper graphs are based. Colony 110: <u>N</u> = 56
workers. Colony 125: <u>N</u> = 52 workers.

Although in Colony 110 defense appears to be concentrated later in this period and in Colony 125 brood care is concentrated earlier, neither pattern holds up in the other colony. These differences between colonies, as well as other differences in frequency distribution among the different roles, are probably influenced by changing social context as the wasps aged. Correlated with the onset of building and other on-nest tasks is the increase in the "visibility" of these workers on the outside of the nest, as seen in Figure 11.8. Thus, workers appear to assume on-nest tasks as they move from inside the nest to spend more time outside, where these tasks are largely performed.

The age at which on-nest roles were initiated, as defined by the onset of building tasks, varied among individual workers. Some began this activity in their first 2-day age interval, but most did so in the 4- to 6-day range (Figure 11.9). Still others never initiated on-nest tasks and were rarely seen outside the nest throughout their lives. The mean ages of onset of building bore no relation to wing length (Table 11.4). There was no correlation between age of first building and age of the switch to foraging in individual workers (Table 11.5).

DISCUSSION

Polybia occidentalis exhibits clear-cut age polyethism. There is clearly a temporal caste specializing in tasks centered at least in part on the outside of the nest. This on-nest caste handles the task-sets of building, nest maintenance, brood care, and defense. As workers age they switch to off-nest tasks, that is, foraging for nectar, water, pulp, and prey. An age polyethism sequence in which the riskiest tasks (foraging) are the last to be taken up yields the highest colony-wide life expectancy compared to other forms of division of labor (Jeanne 1986b).

The evidence for a third temporal caste, an initial one engaging in tasks inside the nest, is indirect. The youngest workers remain inside the

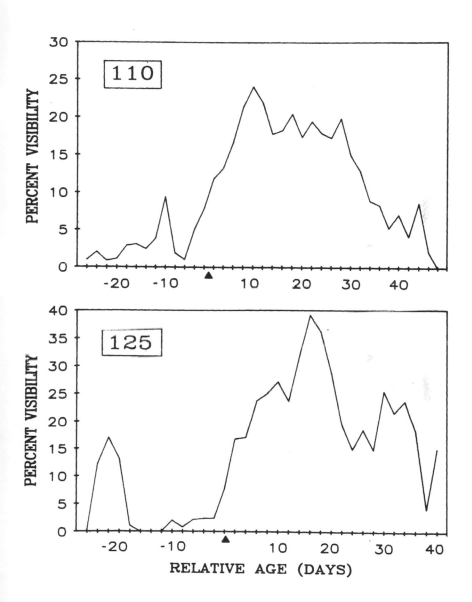

Figure 11.8. Worker visibility. The percent of
    scan samples during which workers were
    visible on the outside of the nest is plotted
    as a function of the age at which each
    individual worker initiated building
    (arrow). Based on scan sample sizes given in
    Figure 11.7. Colony 110: N = 56 workers.
    Colony 125: N = 52 workers.

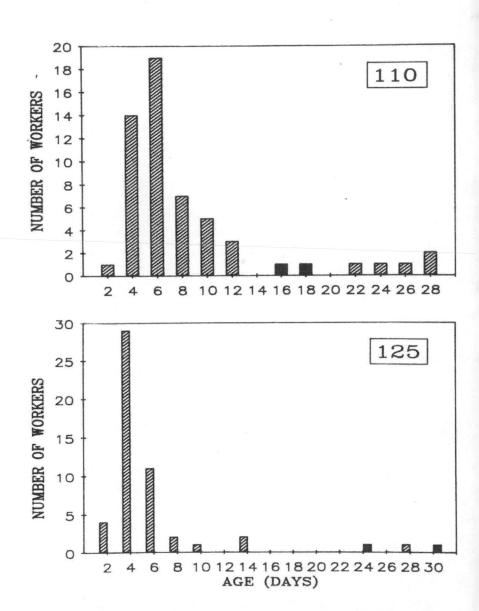

Figure 11.9. Frequency distribution of age of
    initiation of the building role. For the two
    workers in each colony that never initiated
    building (shaded), the age at death is
    given. Colony 110: $\underline{N}$ = 56 workers. Colony
    125: $\underline{N}$ = 52 workers.

Table 11.4
Regression of age of initiation of building on worker body size (wing length)

Colony	N	Age (Days)			Wing Length (mm)			R	t	P
		Mean	SD	Range	Mean	SD	Range			
110	56	7.2	±5.6	1-27	7.5	±0.2	7.1-7.9	0.04	0.28	0.78
125	52	5.1	±5.7	1-29	7.5	±0.2	7.1-8.1	0.05	0.36	0.72

Table 11.5
Regression of age of switch to foraging on age of initiation of building

Colony	N	Age at Start of Building (Days)			Age at Switch to Foraging (Days)			R	t	P
		Mean	SD	Range	Mean	SD	Range			
110	31	6.8	±3.5	4-22	25.8	±7.5	10-40	0.07	0.38	.71
125	37	4.9	±2.2	2-14	19.8	±6.1	8-32	0.28	1.72	.10

nest for several days or more before moving to the outside of the nest to start on-nest tasks. It is not clear what social roles, if any, these youngest workers play. Likely candidates might be brood care and nest maintenance inside the nest. Yet our observations of these task-sets, although extremely scanty and limited to what we could see on the lower comb when it was exposed by removal of the lower envelope, suggest that they are performed contemporaneously with on-nest tasks. On the other hand, two observations suggest that the in-nest phase is not merely a period of idleness during, for example, physiological maturation. First, there is a rather abrupt transition from a period when the workers rarely appear outside the nest to one of frequent appearance there, the latter coinciding with the onset of on-nest tasks. Second, the age at which this transition occurs varies among individuals, with some never leaving the inside. Both phenomena are paralleled in the on-nest/off-nest switch.

At the group level, the switch from on-nest to off-nest roles progresses gradually as the worker population ages, with extensive temporal overlap between the two castes. This result is similar to those of other studies of worker polyethism in social insects. When analyzed at the individual level, however, most of this overlap disappears, revealing a more sudden transition from the one caste to the other. In nearly half the workers we observed no simultaneous performance of the two roles whatsoever. In most others the overlap was limited to a few days. The occasional switching by some individuals between on-nest and off-nest tasks--especially between building and pulp foraging--reported in an earlier study (Jeanne 1986a) to occur during a daily round of nest construction is therefore likely to have been done by workers in this period of transition, rather than by younger on-nest workers or older foragers. That the abruptness of the switch is not due entirely to changing social context--that is, the cessation of building and nest maintenance prior to emigration--is evident from the fact that many workers made the transition either before

preparation for emigration began or after the
colony had initiated the new nest.

Polybia occidentalis appears to have a
"discretized" rather than "continuous" caste
system (in the sense of Wilson 1976). On-nest
task-sets (building, nest maintenance, brood care,
and defense) all have a similar relative age
distribution. The slight differences in
distribution are likely due to rapidly shifting
colony demands as each colony terminated brood
rearing in the old nest, emigrated, and initiated
a new nest.

A colony cycle punctuated seasonally by
emigration, reproductive budding of swarms, and
cessation of production of brood during the dry
season (Forsyth 1978; Jeanne, unpublished data)
means that the colony's worker population probably
never has a chance even to approach a stable age
distribution. Forsyth (1978) concluded that,
because at certain times some age classes will be
missing, age polyethism is not feasible in swarm-
founding wasps. But this is to take the narrow
view that age polyethism means that the transition
from one temporal caste to another is rigidly cued
by absolute age. Clearly this is not the case for
P. occidentalis, and probably does not occur in
any social insect. It seems reasonable to relax
the definition to require merely that the order of
roles through which a worker normally passes as it
ages is fixed. Thus on-nest tasks are performed
before foraging. The length of time spent in each
role may vary among workers, or, in some cases, a
role may be skipped entirely.

The strong influence of season on the colony
cycle of Polybia occidentalis means that the
worker population not only experiences a
continually changing age distribution, but also is
faced with a constantly shifting mix of labor
needs. In this context a system of temporal
castes rigidly tied to absolute age would not
function well. A high degree of flexibility must
be built into the system. How might such a system
work?

It seems necessary to postulate a mechanism
involving age-dependent thresholds of response to
colony needs. The youngest workers would have a
low threshold of response to in-nest tasks. As

they age, first their threshold of response to on-
nest tasks declines, and finally they experience a
decline in threshold for foraging. The age-
dependent threshold profiles would differ among
individuals, some having steeper threshold curves
than others, yielding the wide age overlap of
temporal castes seen at the group level. Unlike
Apis and some species of Bombus, in which larger
workers tend to pass more quickly through the
sequence than do smaller workers (Kerr & Hebling
1964; Free 1955; Waddington 1988), there is no
evidence of such a size correlation in P.
occidentalis. At what age a given worker ceases
performing on-nest tasks, for example, and starts
foraging, would depend on the interaction of its
threshold curve and the relative levels of demand
for the two groups of tasks. The loss of older
foragers to predation, accident, and senescence
creates an elevated demand for foraged materials
to which older nest workers are in a position to
respond by taking up foraging. This effect
cascades its way down to the tasks performed by
the youngest workers, which comprise the ranks of
newly-eclosed recruits to the worker force.

SUMMARY

Based on observations of marked workers of
known age in two colonies in the field, we show
that Polybia occidentalis workers exhibit a
discretized temporal caste system with at least
two temporal castes. The first performs tasks
located primarily on the outside of the nest and
includes the task-sets of brood care, defense,
nest maintenance, and nest building. Most workers
switched from this on-nest role to the off-nest
role of foraging for water, nectar, building
material, and/or prey. Once they had begun
foraging, some individuals showed a tendency to
forage for building material first, then switch to
nectar at an older age; most, however, foraged for
nectar exclusively.

At the group level, the transition from on-
nest to off-nest roles as a function of worker age
occurred gradually. Analysis showed that only 36
and 13 (in the two colonies) percent of the

temporal overlap in roles was due to overlap of
performance by individuals; the remainder was due
to the wide variation in the absolute age at which
individuals made the transition. The age at which
individuals switched roles was not correlated with
body size (wing length).

There was circumstantial evidence for a third
temporal caste, coming ahead of the other two.
During their first few days of adult life workers
were rarely seen outside the nest. Then their
frequency of appearance outside the nest rose
rapidly in conjunction with their taking up on-
nest tasks. The age at which this occurred varied
widely among individuals, suggesting that the
initial in-nest period is not merely one of
physiological maturation.

There was considerable variability among
individuals with regard to which roles were
performed. Approximately two-thirds of the 108
marked individuals used in this analysis
progressed through the "normal" sequence of in-
nest, on-nest, and off-nest phases. Two workers
skipped the on-nest role, and the remainder never
entered the off-nest role, despite most of them
having lived past the mean age for its initiation.

ACKNOWLEDGMENTS

We thank Sean O'Donnell for helping tabulate
data. Brian Yandell provided statistical
advice. Jack Hailman suggested ways to improve
the manuscript. Research supported by the College
of Agricultural and Life Sciences, University of
Wisconsin, and by NSF grant BNS-8112744 to R. L.
Jeanne.

REFERENCES

Brian, M.V. 1979. Caste differentiation and
    division of labor. In Social insects, ed.
    H.R. Hermann, Vol. 1, 121-222. New York:
    Academic Press.
Butler, C. 1609. The feminine monarchie. On a

treatise concerning bees, and the due
ordering of them. Oxford: Joseph Barnes.
Forsyth, A.B. 1978. Studies on the behavioral
ecology of polygynous social wasps. Ph.D.
diss., Department of Biology, Harvard
University, Cambridge, Massachusetts.
Free, J.B. 1955. The division of labour within
bumblebee colonies. Insectes Sociaux 2:195-
212.
Hunt, J.H., Jeanne, R.L., Baker, I., and Grogan,
D. E. 1987. Nutrient dynamics of an
advanced social wasp species, Polybia
occidentalis (Hymenoptera: Vespidae).
Ethology. 75:291-305.
Jeanne, R.L. 1986a. The organization of work in
Polybia occidentalis: costs and benefits of
specialization in a social wasp. Behavioral
Ecology and Sociobiology 19:333-341.
Jeanne, R.L. 1986b. The evolution of the
organization of work in social insects.
Monitore Zoologico Italiano (N.S.) 20:119-
133.
Kerr, W.E., and Hebling, N.J. 1964. Influence of
the weight of worker bees on division of
labor. Evolution 18:267-270.
Naumann, M.G. 1970. The nesting behavior of
Protopolybia pumila in Panama (Hymenoptera:
Vespidae). Ph.D. diss., Department of
Entomology, University of Kansas, Lawrence,
Kansas.
Oster, G.F., and Wilson, E.O. 1978. Caste and
ecology in the social insects. Princeton,
New Jersey: Princeton University Press.
Seeley, T.D. 1982. Adaptive significance of the
age polyethism schedule in honeybee
colonies. Behavioral Ecology and
Sociobiology 11:287-293.
Simões, D. 1977. Etologia e diferenciação de
casta em algumas vespas sociais (Hymenoptera,
Vespidae). Ph.D. diss., Department of
Biology, University of São Paulo, Ribeirão
Preto, SP, Brazil.
Waddington, K.D. 1988. Body size, individual
behavior, and social behavior in honey
bees. In Interindividual behavioral
variability in social insects, ed. R.L.
Jeanne, 385-417. Boulder, Colorado: Westview

Press.
Wilson, E.O.   1971.   The insect societies.
Cambridge, Massachusetts: Harvard University
Press.
Wilson, E.O.   1976.   Behavioral discretization and
the number of castes in an ant species.
Behavioral Ecology and Sociobiology 1:141-
154.
Wilson, E.O.   1980.   Caste and division of labor
in leaf-cutter ants (Hymenoptera: Formicidae:
Atta).   I. The overall pattern in A.
sexdens.   Behavioral Ecology and Sociobiology
7:143-156.

12

Undertaker Specialists
in Honey Bee Colonies

P. Kirk Visscher

Cecropias innatus apes amor urget
habendi munere quamque suo
An inborn love of possession impels the bees
each to his own office (Virgil, Georgics 4)

"Undertaker" bees, as the name suggests, are
those individuals within the honey bee society
which participate in the disposal of their dead
nestmates. This dead-carrying or "necrophoric"
behavior (Wilson et al. 1958) is of particular
interest as an adaptation to one of the
disadvantages of social living. Sociality in
insects is commonly thought of as a breakthrough
allowing a species to gain the best of both the
small world of the individual insect and the
larger world of the superorganism-like colony.
However, there is a dark side to urban
civilization, and insects require a constellation
of adaptations which are corequisites to taking up
life in large colonies.
    The genus Apis consists of five species
(Ruttner 1968; Sakagami et al. 1980). Three of
these are limited to southern Asia and nest on a
single exposed comb hanging from the branch of a
tree or a cliff. The other two species nest
inside cavities in trees or rocks, and have much
more extensive ranges, extending into regions of
cold winters, where the ancestral exposed-comb
nest provides insufficient protection. Adoption
of cavities for their nests required that honey
bees remove any detritus, including dead colony

members, which, in open nests, would simply fall
to the forest floor. If not removed, these
corpses would contribute to the spread of
diseases, and eventually even fill up the nest
cavity.

So bees remove their dead from the nest, and
anyone who watches the entrance to a honey bee
colony for some time will see a worker bee
struggling to drag a dead nestmate to the entrance
and slowly take to the air with this burden and
fly away. These "bee funerals" have provoked
poetic flights of fancy among bee-watchers, such
as the following from Richard Hoy in 1788.

> About twenty-four bees...divide on the way
> coming out, to make way for the dead
> bees....[These] are brought out...after their
> guards are placed in their proper
> station...to keep the way clear. Then comes
> the corpse, carried by four bees to the end
> of the landing board, followed by the
> mourners; and when all is silent, two of the
> strongest bees take hold of the body, lift it
> up, and fly off with it...leaving all the
> rest of the mourners and guards waiting for
> their return. Then the bees which have
> buried dead return first into the hive,
> followed by all the guards and mourners. The
> working bees, taking all pains to collect
> provision for the Winter Store, pass backward
> and foreward all the while, without taking
> the least notice of the funeral.

More objectively, Suzuki et al. (1974)
described the removal of corpses placed just
outside a hive, and Morse (1972) reported that
within the nest a corpse is moved in increments by
successive bees until it is finally taken out the
nest entrance. Attfield (1972) demonstrated that
even living bees are removed if they cannot move,
but Esch's (1964) observations on the treatment of
models and bees within a hive suggested that
chemical cues were also involved in eliciting
necrophoric behavior. Necrophoric behavior in
ants appears to involve a chemical signal,
apparently from fatty acids (Wilson et al. 1958;
Blum 1970; Howard and Tschinkel 1976), but an

ant's behavioral response to this signal depends
upon the social context in which it is encountered
(Gordon 1983).

Little was known about how this behavior fit
into the complex social behavior of honey bees.
This paper, based on Visscher (1983), reports an
investigation of the mechanisms releasing
necrophoric behavior and of the social
organization of this behavior.

DESCRIPTION OF NECROPHORIC BEHAVIOR

Magnitude of Problem

To quantify the significance, or at least the
magnitude, of necrophoric behavior within the
honey bee colony, the dead bees that were carried
out of the hive each day were collected using dead
bee traps constructed according to Gary's (1960)
design (Figure 12.1). These traps, attached to
the entrance of each colony, prevent nest-cleaning
bees from flying away from the hive with dead
bees; when laden with a corpse they are unable to
fly at the steep angle required to leave the
trap. A layer of petrolatum on the underside of
the rim prevents them from crawling out of the
trap. Corpses eventually are dropped and remain
in jars attached to the trap.

The recovery rate of the dead bee traps was
calibrated by introducing 100 marked dead bees
into each hive and noting the number of marked
bees recovered the next day in the traps. On 12
July 1977 this procedure yielded a 96-99 percent
recovery rate. However, when the procedure was
repeated on 13 September 1977 the recovery rate
was only 4 percent. This difference resulted from
an increase in the number of social wasp workers
(Vespula maculifrons Buysson) foraging on the
corpses in the traps in September. During July
1977, 76 counts of corpses in the dead bee traps
on 5 colonies were taken (Figure 12.2), and during
this period wasp foraging was not intense enough
to depress the recovery rates. Each day 54 ±9
(mean ±95 percent Confidence Interval, range 6-
255) dead bees were removed from each colony.

362

Some living bees, especially those which could fly
poorly or not at all, were prevented from leaving
and ended up in the jars attached to the trap.
The above counts do not include those which were
alive when I emptied the traps, but some certainly
had died and were counted. The volume of corpses
totals about a liter a month; it is a fraction of
all the colony's bees which die each day, but does
pose a formidable health hazard.

Necrophoric Behavior Within the Nest

In order to understand the behavior of the
undertaker bees, I initially observed the
interactions of bees in an observation hive with

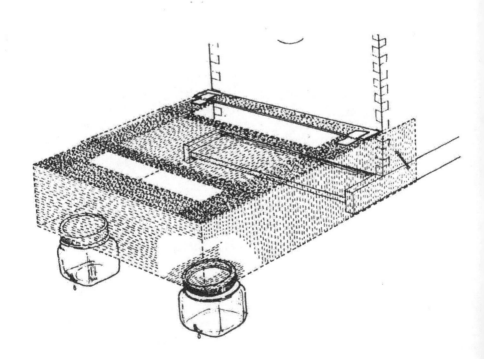

Figure 12.1. Trap for recovering dead bees
    removed from a colony (from Gary 1960).
    Corpses can be collected from the jars where
    they are abandoned by undertakers.

corpses. Bees killed by freezing and then
rewarmed were introduced individually into the
bottom space of an observation hive. This space
was occupied by worker bees moving in and out of
the hive or from one side of the comb to
another. The time required for dead bees to be
removed is extremely variable, and is affected by
such variables as weather, crowdedness of the
colony, and the location of the corpse within the
hive. Bees were carried out very slowly or not at
all when it rained or the temperature was low.
Removal was slower in less crowded colonies, in
which the bottom board of the hive was less busy,
or when the corpse was far from the entrance,
where there were few bees moving about.
     When a bee encountered the corpse, she did
one of the following (percentages based on
observation of 77 interactions of living workers
with five dead bees):

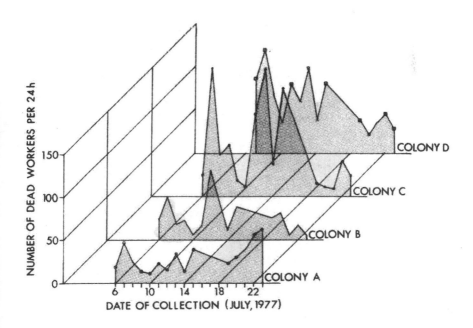

Figure 12.2. The number of dead bees removed each
     day from each of four colonies.

1. 10.4 percent encountered the corpse in their path but seemed to ignore it completely, turning aside or simply walking over it.
2. 36.4 percent encountered the corpse, antennated it for less than 1 second, and then either turned away or walked over the corpse.
3. 6.5 percent encountered the corpse briefly, drew away to about 1 cm, and then reapproached the corpse briefly.
4. 31.2 percent licked the corpse with the proboscis or antennated it closely for more than 1 second.
5. 9.1 percent grasped the corpse in the mandibles and moved it for 1-5 cm, or pulled on it for 1-30 seconds.
6. 6.5 percent grasped the corpse and moved it for more than 5 cm, or pulled on it for more than 30 seconds.

The behavior of the bee that effected the removal of a corpse from the hive seemed more purposeful and rapid than that of bees whose inspections did not result in removal. After a brief inspection, this bee seized an appendage of the corpse in her mandibles and carried it away.

In the observation hive, the route taken by bees carrying corpses was surprisingly circuitous. Figure 12.3 shows the paths of removal of five corpses introduced into a one-frame observation hive through a hole above the top back of the frame. Obviously, it is possible that some disorientation of the bees was caused by the unusual geometry and transparent walls of the observation hive. To evaluate the latter variable, I repeated the experiment with opaque covers on the hive. While the route taken could not be known, the mean time to removal for 5 corpses did not decrease in the covered hive (P >.06, one-sided t test).

Diurnal Pattern of Necrophoresis

I monitored the daily pattern of corpse removal by periodically emptying the traps throughout the day and night. Dead individuals

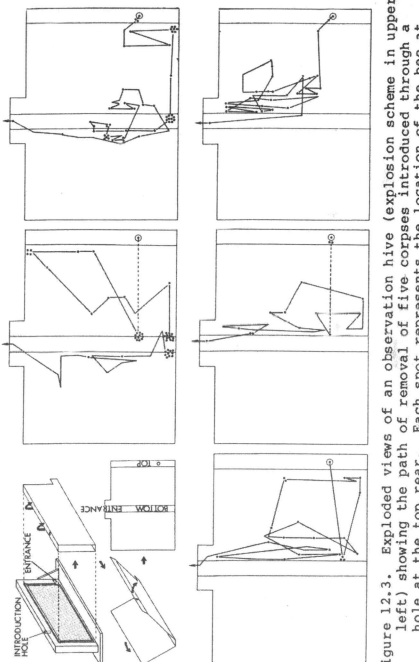

Figure 12.3. Exploded views of an observation hive (explosion scheme in upper left) showing the path of removal of five corpses introduced through a hole at the top rear. Each spot represents the location of the bee at the ends of successive 30-second intervals.

were removed from the hive by the bees throughout the night. This is surprising, since normally necrophoric behavior involves flight (but the traps prevented undertakers from flying in these observations). Figure 12.4 shows the diurnal pattern observed for four hives in Massachusetts in July 1977 and for four hives in Florida on New Year's Eve 1979-80.

Necrophoric Behavior Outside the Nest

Unlike ants, honey bees do not have a refuse pile or other specific locale to which corpses are taken (there are bee undertakers, but there is no bee cemetery). What factors, then, are involved in the eventual abandonment of the corpses? The consummation of necrophoric behavior may require some time out of the nest or removal of the corpse some distance. Normally the undertaker bee will

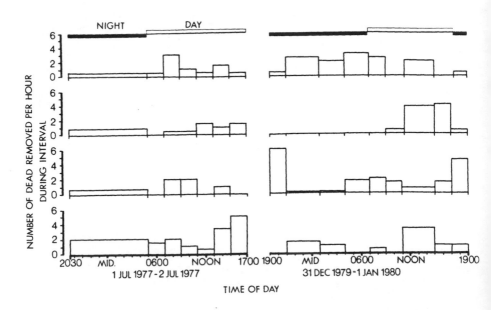

Figure 12.4. Patterns of removal of dead bees through 24 h from undisturbed colonies. The width of bars reflects the length of sampling intervals. The line above the histograms represents the times of sunset and sunrise.

fly from the nest entrance with the corpse grasped in her mandibles, but it is also common for the undertaker and corpse to fall to the ground from the entrance, where the corpse becomes entangled in grass and is sometimes abandoned. Corpse-bearing bees do not readily abandon corpses in the Gary traps, but carry them around for some time before dropping them, and do not let go even when the corpse is seized with forceps.

In order to quantify removal distances, I marked bee corpses with fluorescent paint and introduced them to a hive at the edge of a large paved area, where I searched for the corpses at night with an ultraviolet lamp. The paint used, however, altered the behavior and a greater than normal fraction of the corpses were simply dropped out of the nest entrance. I did recover marked corpses up to 100 m from the hive, and, in this and other observations using unmarked corpses, nearly all of the flights of undertaker bees are further than 10 m. There is some risk involved in these long flights; on two occasions I saw the large, slow-flying pair (corpse and bearer) eaten in mid-air by a scrub jay (Aphelocoma coerulescens). The dead bees are probably eaten by scavenging insects fairly soon after they are dropped by the undertaker bees. I have observed ants feeding on dead bees on the ground as well as the intense foraging by social wasps mentioned above.

MECHANISMS RELEASING NECROPHORIC BEHAVIOR

Bioassay

To investigate the cues involved in releasing necrophoric behavior in honey bees, I developed an assay of necrophoric activity which involved introducing samples of corpses or models with different treatments and comparing their removal times.

"Freshly killed bees" (FKB) were obtained by shaking live bees from combs, narcotizing them in carbon dioxide, and then freezing them in liquid nitrogen. Corpses were then kept frozen on dry ice; prior to use they were thawed and brought to beehive temperature by incubation for 20 min at 37°C. I then introduced them into the beehive by lifting the upper hive boxes to expose the bottom box and scattering the sample between the combs in the center rear of this hive box. I timed the removal of each marked corpse with a stopwatch from its introduction until I seized it with forceps as it was carried out of the hive entrance or as it was carried around in the dead bee trap attached to the hive.

Analysis of Truncated, Non-normal Data

Because the removal-time data obtained from this assay were not normally distributed, I transformed the removal times to their square roots, which were distributed very nearly normally (based on the linearity of probit plots). In many cases, not all samples introduced were recovered in the observation period. In order to compare these truncated samples, I fitted a normal distribution to all of the samples with an iterative procedure (D.S. Robson, personal communication), as shown in Figure 12.5. Differences between samples were then analyzed with a t-test. Because of the variability of the speed of necrophoresis from day to day, only samples introduced to the same hive at the same time may be compared.

Marking Bees

The behavioral assays for the mechanism releasing necrophoric behavior required that several groups be distinctively marked. Typewriter correction fluid (Liquid Paper®), which dries very rapidly, serves this purpose well. As shown in Table 12.1, tests 1a and 1b, it does not affect the removal times of FKB (P = .927).

Necrophoresis Versus Nest Cleaning

It is clear that dead bees in a honey bee

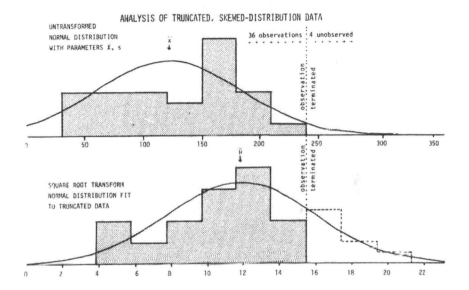

Figure 12.5. Truncated, non-normal data (corpse removal time in minutes) resulting from the necrophoresis assay (top) transformed to square root, and a normal distribution fitted to the resulting near-normal truncated distribution (below). This distribution can then be used in parametric tests.

colony are distinguished from living bees and are removed. Other movable objects are also removed, however, making it necessary to distinguish between truly necrophoric behavior and other nest cleaning. To establish whether necrophoric behavior is distinguished from general nest-cleaning behavior, I introduced dead bees (FKB) and wooden models of bees to a colony. The models were of balsa wood with wire appendages, the same shape and weight as a dead bee, and presumably as easy to grasp. As shown in tests 2a and 2b in Table 12.1, the dead bees were removed much more rapidly ($\underline{P}$ <.001) than the wooden models.

Dead bees, then, elicit an especially rapid response: they are not simply treated in the same way as motionless foreign objects.

Table 12.1
Removal times of dead bees

Test No.	Treatment	N^a	R^b	M^c	γ^d	σ^e	p^f
1 a	FKB + cut wing	10	10	53.4	7.08	2.92	control
b	FKB + Liquid Paper	10	9	58.0	7.18	1.64	.927, NS
2 a	FKB	20	20	7.2	2.85	11.01	control
b	wooden model	14	4	>450	16.72	6.51	<.001
c	extracted	20	16	14.6	6.37	5.14	.017
3 a	FKB	10	10	3.4	1.94	0.65	control
b	extracted	10	9	9.2	3.12	1.05	.013
c	paraffin-coated	10	10	8.0	3.17	1.26	.017
4 a	FKB	40	36	85.1	9.23	2.09	control
b	incubated 12 min	20	20	58.9	7.68	2.51	.013
c	incubated 26 min	20	19	70.3	8.38	2.39	.093, NS
d	incubated 58 min	20	20	42.8	6.54	3.08	.001
e	incubated 108 min	20	19	57.0	7.55	2.50	.009
5 a	extracted + solvent	20	20	40.8	6.39	2.03	control
b	extracted + extract	20	20	27.8	5.27	2.40	.060, NS

[a] Number of dead bees in each sample introduced.

[b] Number recovered.

[c] Median of removal time in minutes.

[d] Mean of the normal distribution fitted to the square root transforms of removal times. Note that $(\mu)^2 = M$, especially for the larger samples.

[e] Standard deviation of the normal distribution fitted to the square root transforms of the removal times.

[f] Probability under the null hypothesis of no difference between each treatment and its control (t-test, two-sided for tests 1-3, one-sided for tests 4-5, which had a priori hypotheses).

Honey bee corpses are usually removed from
the hive in less than an hour, whereas bits of
wood or paper usually require more than an hour.
There are numerous ways in which dead bees differ
from living bees or debris, but which are used in
˙istinguishing a dead bee within the nest?

It appears from the behavior of nest-cleaning
bees and the experiments with models that some
close-range, possibly contact cues are used to
identify dead bees. Some sensory channels can be
dismissed from consideration as bearing signals of
"deadness". Vision is not involved, as
necrophoresis takes place inside dark nest
cavities, dark areas of beehives, and observation
hives illuminated through red filters admitting
only light beyond the spectral sensitivity of
bees. Neither auditory nor thermal cues
distinguish corpses from other foreign objects.
Tactile and chemical cues remain.

Tests for Chemical Releasers

To test whether bees use a chemical cue to
recognize their dead nestmates, I compared FKB to
bees coated with paraffin and to bees extracted
for 90 min in each of five solvents: hexane,
ether, acetonitrile, methanol, and water.
Extractions were made in a Soxhlet extractor,
which washes huge volumes of pure solvent over the
bees but concentrates the products in a small
solvent reservoir (provided they are less volatile
than the solvent and stable at its boiling point).

As shown in Table 12.1, test 2c versus 2a,
and tests 3b and 3c versus 3a, bees which have
been chemically extracted or coated with paraffin
are removed more slowly than untreated dead bees
(P <.02). These results imply a chemical signal,
although each of these treatments also alters the
tactile cues of a corpse.

If the signal is chemical, it probably begins
to develop after death, as suggested for ants
(Wilson et al. 1958; Blum 1970; Howard and
Tschinkel 1976). To test this, FKB were incubated
for 12, 26, 58, and 108 min at 37°C and refrozen
in liquid nitrogen. These samples and control FKB
were then incubated for an additional 20 min and

then introduced to a hive. As shown in Table
12.1, test 4, bees incubated 12 min longer than
controls were removed more rapidly, but longer
incubation did not result in any further
significant decrease in removal time.

Freshly killed bees and corpses dead for some
time are removed at different rates, but do not
differ in texture, and there is no obvious
difference in rigidity between the corpses after
different treatments. This, and the observation
that extraction of dead bees diminishes the
necrophoric response they elicit, both suggest
that the releaser of necrophoric behavior is
chemical. As a further test of this hypothesis,
FKB were incubated at 37°C for 30 min and then
extracted in cold chloroform for 60 min. I then
added 0.05 cm^3 of the chloroform extract per bee
to a group of 20 FKB, and added 0.05 cm^3 of
solvent to each of 20 control FKB. The bees were
incubated 30 min at 65°C to evaporate the solvent,
and then introduced to the hive. The results are
presented as test 5 in Table 12.1. The median
removal time for the treated corpses was shorter,
but this difference only bordered significance
($\underline{P}$ = .06). Only 0.23 bee equivalents of the
extract were readded, however, and some of the
activity could have been lost in the extraction or
heating.

Changes in a bee after death, probably
initiated at death, create a signal remarkably
rapidly. Whether this is due to a loss of odors
inhibiting necrophoresis or to the appearance of
odors releasing it, or both, I did not
determine. However, the decrease in removal time
due to adding dead bee extracts to extracted bees
argues for a releaser effect of chemicals
appearing after death, or at least a critical
balance between 'liveness' and 'deadness'
smells. Probably necrophoric behavior in honey
bees is not as simple as being mechanistically
released by a single chemical, and involves some
recognition that the object encountered is a bee,
and further recognition that it is a dead one.
Wilson et al. (1958) reported that <u>Pogonomyrmex</u>
ants would carry living workers painted with oleic
acid to the refuse heap, but Gordon's (1983)
results suggest that here, too, the recognition is

more complex than that of a simple releaser model
and is dependent on the behavioral state of the
ants involved.

SOCIAL ORGANIZATION OF NECROPHORIC BEHAVIOR

Specialization of Undertakers

The seemingly purposeful behavior of the bees
which effected necrophoresis suggested that they
might be specialized to some extent in this
task. To test this hypothesis, I introduced dead
bees to a small colony and then marked each bee
which I saw removing a dead bee from the hive
entrance. To mark undertaker bees as they left
the colony while not disturbing their subsequent
behavior and that of the colony's foragers, I used
the marking trap depicted in Figure 12.6. Like
the dead bee trap described earlier, this trap
allows unencumbered bees to come and go freely
through the upper opening. Undertaker bees cannot
take off steeply enough to fly through this
opening, and must drag their burdens through the
lower tunnel. The mesh forming the top of this
tunnel is flexible and bowed upward. When the
wire harness attached to it is pulled from below,
the tunnel roof flattens and pins any bees in the
tunnel in place. They can then be marked through
the mesh. In most cases, when the harness is
released the marked bees continue normally with
their activities.

When a bee I observed and marked as an
undertaker later carried out another dead bee, I
added another mark of a different color so that I
had a record of the number of corpse removals for
each of the bees (Table 12.2). I performed this
experiment in January 1980 with a one-frame
observation colony and in January 1981 with a
three-frame observation colony. I estimated the
population of these colonies by three counts of
randomly chosen squares of a grid covering both
sides of the hives.

Table 12.2 indicates that there is clearly a
group of undertaker bees specialized in
necrophoric behavior, since many more multiple

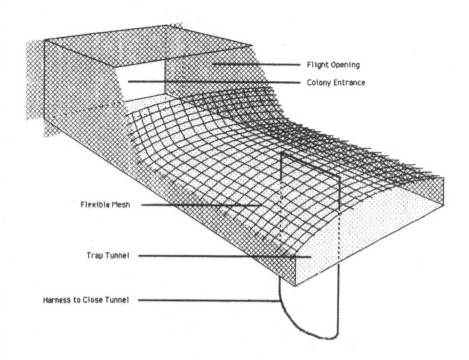

Figure 12.6. Trap for marking undertaker bees as they remove corpses. The body of the trap is constructed of hardware cloth and the tunnel roof of flexible wire mesh.

removals were observed than expected if all bees performed this work. Using the methods of the Schnabel census in mark-recapture studies (Seber 1973) it is possible to calculate a maximum-likelihood estimate for the size of this group of undertaker bees from these data. Different 'batch' sizes, that is, the number of times an individual might possibly have removed a corpse, yield different estimates. Since batch size could range from the largest number of corpse removals observed for any one bee to the total number of corpses removed, I calculated an estimate of the number of undertaker bees for each of these extremes of batch size (Table 12.2).

About one or two percent of the bees in a colony appear to specialize in corpse removal.

Table 12.2
Specialization of undertaker bees

	1980	1981
Bees removing		
1 corpse	9	60
2 corpses	15	21
3 corpses	5	6
4 corpses	2	2
5 corpses	0	2
6 corpses	0	1
Total	31	92
Estimated number of undertaker specialists[a]	32–38	131–145
Hive population[b]	3640 ±110	6139 ±477
Percent of undertakers	0.9–1.0	2.1–2.4

[a]The range in estimates of number of specialists is due to analysis using two extreme possible cases (see text).

[b]Hive populations are mean and SD of three estimates.

The 1981 data suggest that within this group not all bees are equally likely to remove a given corpse: some may be more specialized or more active.

Age and Tenure of Undertaker Bees

Figure 12.7 shows the age distribution of the bees Sakagami (1953) observed removing corpses. Undertaker behavior occurs after the individuals involved have taken their first orientation flights (as would be expected, since these bees must fly away from and back to the hive), but before they begin foraging. This overlaps with the period of guarding behavior, and indeed I saw a few of the undertaker bees I had marked in my experiments acting as guard bees on the same day that they removed corpses.

Individuals apparently remain undertakers for at least several days. Sakagami (1953) reported that out of fifteen bees seen removing corpses, three were seen doing so more than once (Figure 12.7), and these removals were 2, 6, and 7 days apart (since his study did not concentrate on undertakers, his methods resulted in observing

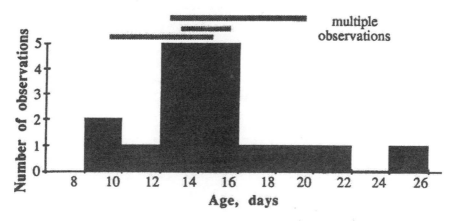

Figure 12.7. Age distribution of undertaker bees. Data from Sakagami (1953). Age span of bees seen removing corpses more than once is shown in the lines above the histogram.

necrophoresis only rarely, so his figures on tenure can serve only as minima). This relatively long tenure as undertakers accentuates the specialization of these bees, since a group of undertaker specialists of a given size represents a smaller percentage of the bees in their age class when that age class is broader.

This idea can be generalized for castes in social insect colonies. It would not apply to genetically or morphologically determined castes, but would apply to subcastes within these. The proportion of individuals that ever participate in a caste can be represented:

$$\frac{\text{Proportion ever serving in a given caste}}{} = \frac{[\text{Average number (M) of individuals in caste at given time}] \times [\text{Average number of days of life (L)}]}{(\text{Size of colony, N}) \times (\text{Days tenure in caste, T})}$$

The variables in this equation have been estimated for undertaker bees, with the exception of T. Using T = 5 as a guessed minimum from Sakagami's data above, L = 33 (Ribbands 1953) and M/N = 0.015 from the data of this study yields a maximum proportion of 10 percent of bees in a colony which serve as undertakers at any time in their life.

The small specialized force of undertakers, which comprise 1-2 percent of the colony population, may be the bees which are most likely to approach corpses in the hive and inspect and remove them quickly. However, the circuitous path of removal of corpses, often involving several bees, suggests that the specialized bees that finally remove corpses may not always be the bees that carry these corpses from deep within the nest.

CONCLUSIONS

Undertaker specialists are of particular interest in the study of division of labor in insect societies in part because they provide an example of extreme specialization on a single task

by a very small subset of a colony. Nearly all
other studies on honey bee division of labor have
concentrated on the large-scale patterns of
division of labor (reviewed by Seeley 1982, 1985;
Wilson 1971; Ribbands 1953). Probably undertaking
is just one example of a much more widespread
pattern of narrow specialization by individuals on
specific tasks. Lindauer (1953) demonstrated that
relatively few bees ever serve as guards for the
colony, but those that do so remain guards for a
long period. Specialization in foraging by
individual workers is well known [reviewed by
Ribbands (1953) and Free (1970)], ranging from
simple flower constancy on foraging trips to
Robinson et al.'s (1984) report of an individual
bee's specialization on water collection for her
entire foraging life.

Undertaker bees within a colony, specializing
as they do in the task of corpse removal, comprise
a caste in the broad sense (Oster & Wilson
1978). To call undertakers a caste, however, may
be stretching that definition beyond its useful
limits. On the other hand, it is important to
recognize that rare behaviors such as this may
have their specialists, and that due to their
rarity they tend to be overlooked in studies with
a coarser grain. To identify every group of
specialists in a colony as comprising a caste has
two drawbacks. First, it obscures important
patterns revealed by broader scale definitions in
a forest of detail. Second, at the fine level
there is the possibility of considerable overlap
in behavioral specializations, so that an
individual might serve as both an undertaker and a
guard, or as an undertaker and a food-handler.
While this problem also occurs in the extremely
broad temporal caste definitions suggested by
Seeley (1982), in that case it is principally a
result of transition from one caste to another.

Narrow specialization within the broader
pattern of temporal castes may prove to be common
in honey bees and other social insects, although
the unusually well-defined nature of the task of
necrophoresis makes specialization here easier to
detect. It may also make specialization more
likely. Undertaking is a complicated task, and it
is likely that an undertaker bee learns how to do

it well with practice, but it is a task that is quite rare in a colony and which is not directly connected with other activities a bee might be performing. In Seeley's (1982) terms, it is likely that this sort of task is one on which a high degree of specialization can lead to a high degree of performance efficiency. Task location efficiency may be low if the task is rare, although the predictability of where corpses are found (they fall) makes it locally less rare. Finally, corpse removal may be an especially critical behavior for the colony, putting a premium on performance efficiency and on speed, rather than efficiency of task location.

The above characteristic also makes the undertaker bee system attractive for experimental work. Here we have an uncommon, distinctive behavior performed by specialized individuals, and one which can be experimentally manipulated and observed quite easily. Necrophoresis can be elicited on demand by simply introducing dead bees to the colony, and this manipulation probably has very few side effects on the dynamics of the colony, unlike manipulation of forage or brood. These attributes make the undertaker bee system ideal for asking what factors regulate the allocation of labor to specific tasks in the colony.

For example, one could provide two sets of hives with different numbers of dead bees to be removed to see whether the frequency of encounters with appropriate tasks, here corpses to remove, affects the tenure of individual bees in a specific role in the labor of the colony, or whether it affects the probability of a given bee ever taking up the role of undertaker.

In conclusion, it appears that a true necrophoric behavior is part of the honey bee's repertoire of nest hygiene behavior. It is probably released by a chemical that is perceived on contact or within a small active space and which appears in honey bee corpses shortly after death. The bees that die in the nest each day would typically accumulate at a rate of about one liter per month, filling the nest cavity and probably contributing to the spread of diseases and parasites within the colony. This

accumulation of dead colony members is an example
of one of the disadvantages of social living in
enclosed nests, a problem largely overcome by the
mechanisms of necrophoric behavior in honey bee
colonies. The study of undertaker bees adds a new
detail to our knowledge of honey bee behavior, is
an instructive example of extreme specialization
in colony labor, and provides an attractive
experimental system for answering some of the
broader questions of the mechanisms of labor
allocation in insect societies.

SUMMARY

Particular bees within a colony of honey bees
(Apis mellifera) specialize on the removal of bees
which die within the nest. Because of the health
hazard that corpses represent within enclosed
nests, this behavior is an essential adaptation in
the social evolution of honey bees. Only 1-2
percent of the colony members participate in
removal of the dead at a given time, and only
about 10 percent of all bees ever participate.
Corpses are removed more rapidly than other debris
in a colony; the cues releasing undertaker
behavior are probably chemical, and develop
rapidly after the death of a bee.

REFERENCES

Attfield, H.H.D. 1972. Treatment of injured
    worker bees in a colony of Apis mellifera.
    American Bee Journal 112:52.
Blum, M.S. 1970. The chemical basis of insect
    sociality. In Chemicals controlling insect
    behavior, ed. M. Beroza, 61-94. New York:
    Academic Press.
Esch, H. 1964. Beiträge zum Problem der
    Entfernungsweisung in den Schwänzetänzen der
    Honigbienen. Zeitschrift für Vergleichende
    Physiologie 48:534-546.
Free, J.B. 1970. Insect pollination of crops.
    London: Academic Press.

Gary, N.E. 1960. A trap to quantitatively recover dead and abnormal honey bees from the hive. Journal of Economic Entomology 53:782–785.

Gordon, D.M. 1983. Dependence of necrophoric response to oleic acid on social context in the ant, Pogonomyrmex badius. Journal of Chemical Ecology 9:105–111.

Howard, D.F., and Tschinkel, W.R. 1976. Aspects of necrophoric behavior in the red imported fire ant, Solenopsis invicta. Behaviour 56:157–180.

Hoy, R. 1788. Directions for managing bees in Hoy's octagon box beehives. London: Hoy's Honey Warehouse.

Lindauer, M. 1952. Ein Beitrag zur Frage der Arbeitsteilung im Bienenstaat. Zeitschrift für Vergleichende Physiologie 34:299–345.

Morse, R.A. 1972. Environmental control in the beehive. Scientific American 226:93–98.

Oster, G.F., and Wilson, E.O.. 1978. Caste and ecology in the social insects. Princeton, New Jersey: Princeton University Press.

Ribbands, C.R. 1953. Behavior and social life of honeybees. London: Bee Research Association.

Robinson, G.E., Underwood, B.A., and Henderson, C.H. 1984. A highly-specialized water-collecting honey bee. Apidologie 15:355–358.

Ruttner, F. 1968. Systématique du genre Apis. In Traité de biologie de l'abielle. Vol. 1, ed. R. Chauvin, 2–26. Paris: Masson.

Sakagami, S.F. 1953. Untersuchung über die Arbeitsteilung in einem Zwergvolk der Honigbiene. Beiträge zur Biologie des Bienenvolkes, Apis mellifera L., I. Japanese Journal of Zoology 11:117–185.

Sakagami, S.F., Matsumuro, T., and Ito, K. 1980. Apis laboriosa in Himalaya, h ittle known world largest honeybee (Hymenoptera, Apidae). Insecta Matsumurana (New Series) 19:47–77.

Seber, G.A.F. 1973. The estimation of animal abundance and related parameters. London: Griffin.

Seeley, T.D. 1982. Adaptive significance of the age polyethism schedule in honeybee

colonies. Behavioral Ecology and
Sociobiology 11:287-293.
Seeley, T.D. 1985. Honeybee ecology: a study of
adaptation in social life. Princeton, New
Jersey: Princeton University Press.
Suzuki, K., Yoshihama, T., and Shigematsu, Y.
1974. Sweeping behaviours of honey bees at
the hive entrance. Bulletin of the Faculty
of Education, Chiba University 23:273-281.
Visscher, P.K. 1983. The honey bee way of
death: Necrophoric behavior in Apis
mellifera colonies. Animal Behaviour
31:1070-1076.
Wilson, E.O. 1971. The insect societies.
Cambridge, Massachusetts: Harvard University
Press.
Wilson, E.O., Durlace, N.E., and Roth, L.M.
1958. Chemical releasers of necrophoric
behavior in ants. Psyche 65:108-114.

13

Body Size, Individual Behavior, and Social Behavior in Honey Bees

Keith D. Waddington

The essence of social life in insects is the division of labor among coexisting, related individuals. Members of the colony perform tasks that are more or less integrated and the result is passage of genes via reproductives to another generation. Given the numerous tasks that must be completed and the seemingly chaotic setting of the social insect colony, it is a wonder that any genes are passed on. Indeed, much work on social insects has been directed toward making sense of the apparent chaos. Which individuals perform which tasks and how? How are the "components" fit together to make a functional colony?

A close look at the social insects reveals that they have a division of labor based on caste. The hymenopteran colony is comprised of male and female reproductives and a worker caste of sterile females. The workers may be differentiated, depending on species, into subcastes or groups of individuals that perform similar kinds of labor. The individuals within each subcaste have similar morphology and are distinct in morphology from members of other castes (Wilson 1953). Of the Hymenoptera, the ants have the greatest differentiation among the workers. Although workers of most species of ants are monomorphic, some species have two or three morphological subcastes (Oster and Wilson 1978). The morphological forms of these polymorphic species are results of allometric growth and within-colony size variation (Wilson 1971). Where does monomorphism end and polymorphism begin?

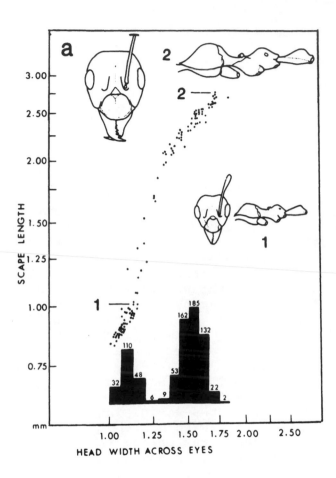

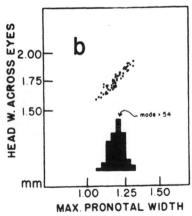

Most authors agree that there must be minimal
detectable allometry (Wilson 1971). The evolution
of caste in ants has tended toward increased
allometry between body parts, general increase in
size variation, and also toward multiple modes in
the size frequency distribution. For example, in
a colony of Oecophylla smaragdina (Fabricius)
there are workers with large or small heads
(called majors or minors), but few workers have
heads of intermediate size (Figure 13.1a). These
two castes perform different tasks. Wilson (1980)
recently studied caste and division of labor in
Atta sexdens (Linné), a leaf-cutter ant. Workers
of this species have high size variation but the
variation is continuous, and there is a single
allometric curve for any pair of morphological
measurements. Although there are no discrete size
classes of workers, ants of different size fall
into four partially discrete physical subcastes
which are based on body size (called "role
clusters"). For example, ants that garden and
nurse tend to be small bodied while the foragers
and excavators are the largest ants of the
colony. Primitive ants such as Formica
exsectoides Forel, in contrast, have workers with
isometric growth and unimodal size frequency
distributions (Figure 13.1b). Workers

Figure 13.1. (a) In Oecophylla smaragdina the
    allometric relationship between scape length
    and head width is triphasic (that is, three
    slopes are shown) and the frequency
    distribution of size is bimodal. The heads,
    mesosomas, and petioles of selected minor and
    major workers are also shown. (b) In the
    primitive formicine ant Formica exsectoides
    the workers are monomorphic and the body parts
    are all isometric with reference to one
    another, as exemplified here by the
    relationship between head width and pronotal
    width (From E.O. Wilson, 1953. The origin and
    evolution of polymorphism in ants. Quarterly
    Review of Biology. The Stony Brook
    Foundation, Inc., pp. 141, 143; used with
    permission.)

of these species are monomorphic and are not considered to have physical subcastes.

Worker behavior in most ants also changes systematically with age; this is referred to as temporal polyethism. For example, young workers of the ant Formica polyctena (Först) stay inside the nest and spend time tending brood, caring for other workers and the queen, and cleaning the nest. Older workers forage and construct the nest (Otto 1958).

Worker bees are generally characterized as monomorphic, thus no physical subcastes are recognized but variation in body size differs among species. Primitively eusocial bees, such as bumble bees (Bombus spp.) and sweat bees (Halictidae), have high within-colony size variation, whereas the highly eusocial honey bees (Apis spp.) and stingless bees (Meliponinae) have low size variation (Michener 1974). As in the ants, worker bees perform a sequence of different tasks as they age. Age polyethism has been particularly well studied in the honey bee (Lindauer 1952; Ribbands 1952; Sekiguchi and Sakagami 1966; Seeley 1982) (Figure 13.2). Seeley (1982) defined four age subcastes: cell cleaners, nest brooders, food storers, and foragers.

The currently used framework for studying division of labor in relation to individual morphology has not encouraged addressing the (not so obvious) question: how is task performance affected by variation in morphology (generally meaning body size) within a monomorphic, isometric worker? I do not know of any study that focuses on variation in task performance of different size ants within a monomorphic species. Similarly, few studies of bees have related body size of workers to behavior. The primary goals of this paper are to characterize some behavior patterns of individual honey bees in relation to body size and to examine the consequences of variation in body size on the functioning of the honey bee colony.

FACTORS THAT AFFECT BODY SIZE

Worker body size is a result of both environmental and genetic factors (Abdellatif

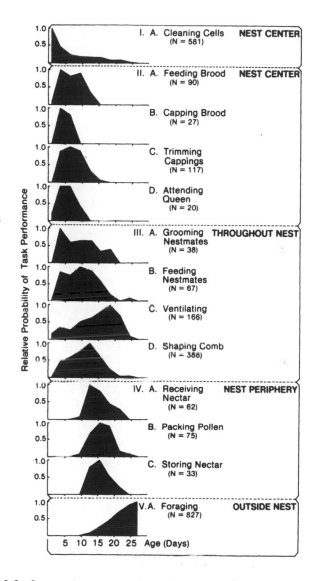

Figure 13.2. The curves of relative probability of task performance for thirteen tasks by Apis mellifera workers of different ages. The curves are classified into five groups which are used in defining the age of castes. (From T.D. Seeley, 1982. Adaptive significance of the age polyethism schedule in honeybee colonies. Behavioral Ecology and Sociobiology. Springer-Verlag, p. 289; used with permission.)

1965). There have been a few studies of the
effects of various environmental factors on size
in honey bees. Cell size affects worker body
weight. Abdellatif (1965) found that bees reared
in brood comb that was seven years old weighed
less than bees reared in newly drawn comb. The
mean volumes of cells of the old and new comb were
0.2446 and 0.2752 cm^3, respectively (Table
13.1). Kulzhinskaya (1956) also found that larvae
reared in larger cells weighed more, as did the
emerged adults, compared with individuals reared
in smaller cells. How cell size actually affects
bee size has not been completely worked out.
Kulzhinskaya (1956) found that worker larvae
reared in larger cells were fed more food and more
protein than larvae reared in smaller cells.

If it is the amount and quality of food that
directly affects adult worker size, then factors
which influence the rate that food can be brought
to the hive may play a role. Potential factors
include food supply, size of the worker force,
weather, and competition for food with flower
visitors from other colonies or other species.
Work is needed in order to understand the
influence of these and other factors on worker
body size. Size differences in bumble bees are
well known to be trophogenic (Plowright and Jay
1977).

Little is known about the genetics of worker
body size; however, a few observations have been
made. Roberts (1956) found differences in size
among inbred lines of workers and Ruttner and
Mackensen (1952) believed that size differences
found among geographical races of bees are due to
additive genetic variance. There is need for work
which examines relationships between within-colony
genetic variation and variation in body size.

In summary, a full understanding of the
significance and ramifications of relationships
between body size and behavior will not be
possible until more is known about genetic and
environmental determinants of body size.

Table 13.1
Grand mean of worker body weight and comb cell
size for four colonies of Apis mellifera. (From
Abdellatif 1965). Reprinted by permission of
the American Bee Journal.

	Grand Mean (±SD)	
Type of Comb	Body Weight (mg)	Cell Size (cm^3)
New comb	109.78 (±0.2240)	0.2752 (±0.0006)
Old comb	100.49 (±0.3390)	0.2446 (±0.0015)

BODY SIZE, FORAGING FREQUENCY, AND IN-HIVE BEHAVIOR

Kerr and Hebling (1964) examined the
relationship between division of labor in a colony
of Apis mellifera adansonii L. ("Africanized
bees") and bee weight. Newly emerged bees were
weighed and marked so that individuals could be
recognized; they were then introduced into a small
observation hive. A number of behavior patterns
were observed and the ages that bees began them
were determined. The experiment was performed
during two different times of the year, summer and
winter. Heavy bees invariably began new tasks
earlier in life than light bees. For example, in
the summer experiment heavy bees began cleaning
cells and foraging on the 1st and 18th day of
adult life, respectively, whereas light bees began
the same tasks at ages 2 and 27 days (Table
13.2). No light bee began a particular task

Table 13.2
Ages at which individually numbered summer bees began certain types of work (From Kerr and Hebling 1964)

Labor	Light bees		Heavy bees	
	Bee Number	Age	Bee Number	Age
1. Cleaning themselves	130	1st day	220	1st day
2. Cleaning cells	156	2nd day	203	1st day
3. Nursing old larvae	156	5th day	203	4th day
6. Nursing young larvae	156	8th day	202	7th day
4. Courtship of the queen	130	7th day	203	4th day
7. Putting food around eggs	110	9th day	202	8th day
8. Collecting honey from a feeding place	156	10th day	204	8th day
9. Receiving nectar	--	--	299	10th day
5. Sealing brood cells	152	15th day	226	4th day
10. Attending communication dance	149	22nd day	205	17th day
12. Cleaning the hive	130	23rd day	201	21st day
11. Field work	110	27th day	205	18th day
13. Communication dance	149	28th day	226	22nd day

before any heavy bee. Light bees used in the
summer experiment ranged from 81 to 88 mg and
heavy bees were between 102 and 110 mg. Kerr and
Hebling (1964:269) concluded that "...weight
is...important in relation to division of labor in
Apis mellifera...."
 As part of a study to investigate the
influence of several factors on honey bee
behavior, Cideciyan (1984) examined the effects of
body size on twelve behavioral acts performed
inside the hive and on foraging behavior. Newly
emerged bees were taken from an incubated comb and
immediately labelled by gluing colored, numbered
plastic tags onto the dorsal thorax. Measurement
D (Figure 13.3) was made on each of 231 bees over
seven days. This measurement was easy to make on
lightly anesthetized bees and was found by
Cideciyan to be positively correlated with other
measurements of body size (Table 13.3); thus, D
was taken as an indicator of overall body size.
The frequency distribution is presented in Figure
13.4.

Foraging Behavior

 Cideciyan (1984) observed bees leaving the
observation hive through a clear plastic tube.
Foraging activity was quantified in two ways: (1)
the number of foraging initiations per observation
period (seventy-four 40-minute observation periods
were made) and (2) trip duration. She found that
body size affected the frequency of foraging
initiations (Kruskal-Wallis test, $H = 14.068$, $N = 91$, $\underline{P} < .05$; dependent variable = mean number of
foraging initiations per bee, independent variable
= body size). Larger foragers tended to forage
more frequently than did smaller bees (Kendall
rank correlation, $t = 0.123$, $N = 91$, $\underline{P} < .05$
[Figure 13.5]). Trip duration did not vary with
body size ($H = 9.028$, $N = 75$, $\underline{P} > .05$).

Behavior Inside the Hive

 Twelve instantaneous behavior patterns were
tallied inside the observation hive (reference for
definition provided in parentheses; see also
definitions in Cideciyan [1984], Table 1): walking

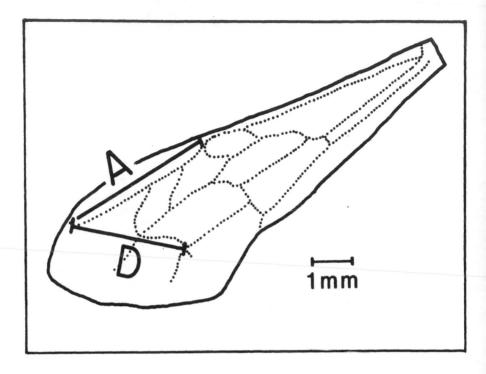

Figure 13.3. Forewing of <u>Apis mellifera</u>.
     Measurement D was used by Cideciyan (1984) as
     an indicator of body size.

(Lindauer 1961:19), resting (Lindauer 1961:19),
grooming (Free 1965:41), queen care (Ribbands
1964:284), inspecting brood cells (Lindauer
1961:16), fanning (Gary 1975:216), cleaning hive
(Gary 1975:218), transmission of food (Gary
1975:213), dancing (Frisch 1967:29, 57), dance
following (Frisch 1967:29, 57), pollen storage
(Lindauer 1961:11), and other (undetermined
activity when head was inside a cell). The
instantaneous activity of a bee was recorded as
the bee was encountered by a wire passed across
the comb. Unfortunately, this method could not be
used to determine patterns of individual behavior
through time.
     Results of this study indicated that body
size affects behavior performed inside the hive
(G-test, G = 101.3, d.f. = 77, $\underline{P}$ <.05). These
data could not be used to discern which tasks were

Table 13.3
Correlation matrix for size indicators (Pearson product-moment correlation coefficients) (After Cideciyan 1984)

	Wing Cell Length	Coxa Length	Tongue Length	Wing Width	Distance between 2 veins
Wing Cell Length[a]	1.000				
Coxa Length	0.519[d]	1.000			
Tongue Length	-0.035 NS	0.055 NS	1.000		
Wing Width	0.882[d]	0.505[d]	0.094 NS	1.000	
Distance between 2 veins[b]	0.739[d]	0.517[d]	0.293[c]	0.794[d]	1.000

[a]Measurement A, Figure 13.3

[b]Measurement D, Figure 13.3

[c]$p < .04$, [d]$p < .001$, NS = $p > .05$.

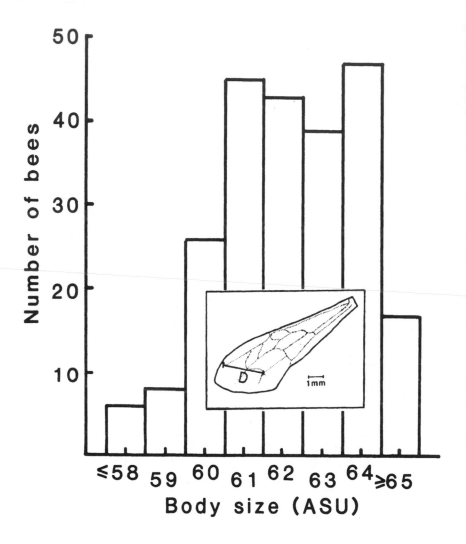

Figure 13.4. Frequency distribution of worker
body size (measurement D, Figure 13.3). One
ASU equals 0.04 mm. (After Cideciyan 1984).

performed differently by different size bees.
Individuals should be monitored continuously in
future studies in order to address this problem
(Cideciyan 1984). As expected (from Lindauer
1961), Cideciyan's study also showed that bees of
different ages performed different tasks inside
the hive.

Summarizing the results of Kerr and Hebling

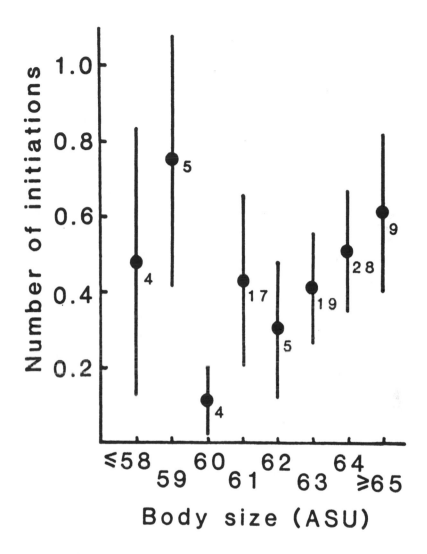

Figure 13.5. Mean (±2 SE) number of foraging initiations per bee calculated for bees of each size (measurement D, Figure 13.3). Ninety-one bees were sampled; number of bees within each size-class is presented.

(1964) and Cideciyan (1984), larger bees in comparison to smaller bees; (1) switch to subsequent tasks (in the normal sequence) at an earlier age (this may explain Cideciyan's "in hive" results), and (2) forage more frequently.

It is still unclear whether 1 and 2 are
independent.

BODY SIZE AND FLIGHT DISTANCE

Economics of foraging depend on decisions as
to which flowers to visit and the behavior
patterns involved in food collection (Waddington
1983a). Also involved is the distance flown
between hive and flowers. Not all foragers of a
colony should visit flowers close by and thereby
minimize flight costs because the food collected
per flower would be low due to intracolony
competition. Bees should disperse over distant
flower patches where profits may be greater.
Presently, little is known about the bees'
abilities to trade off flight distance and
foraging gains. Boch (1956), however, found that
bees do trade off profits and flight distance.
The greater the distance between hive and a feeder
containing sugar solution, the more highly
concentrated the solution must be to elicit
dancing.
In a colony, are payoffs from foraging
affected by the distribution of distances flown to
food by bees of different size? The answer to
this question lies in the economics of foraging by
different size bees and the economics must
differ. As examples, the rate of energy
expenditure (cal/time) during flight is greater
for a larger bee but the speed of flight is also
greater. The carrying capacity, for pollen or
nectar, may be greater for a larger bee (Wells and
Giacchino 1968), but rate of energy expenditure
increases with load-size. However, it is
difficult to know now how the colony's bees should
be distributed as to size because details of the
above relationships are unknown. For now I can
simply report observations on the relationships
between body size and the distance flown for
pollen and nectar (Waddington, unpublished data).
Three hundred and nine newly emerged bees
were measured (measurement D, Figure 13.3) and
individually marked over a one-week period and
were introduced into a glass-sided observation
hive. As marked bees were observed performing the

waggle dance I counted the number of circuits and timed the dance. The number of circuits per unit time was taken as the distance flown between hive and flowers (Frisch 1967). Spearman rank correlation analysis was performed using measurement D (body size) with dance rate (circuits/s). Separate analyses were performed on bees carrying pollen (pollen collectors) and on bees not carrying pollen (nectar collectors). No significant correlation was found between body size and dance rate for nectar collectors (rho = 0.04, $\underline{P}$ >.05, N = 43). However, larger bodied pollen collectors tended to fly further to forage than did smaller bees (rho = 0.75, $\underline{P}$ <.05, N = 10). An explanation of this pattern awaits further investigation. A description of the bees' time and energy budgets and the mass of pollen loads carried over various distances would begin to address this question.

In a pilot study, I did a second investigation of flight distances by different size bees. Bees from a single colony were trained to a feeder placed at three distances from the hive: 195, 595, and 1,161 m. Foragers that danced inside the hive were sampled and wing measurement D was made on each. The mean (±1 SD) value of D of the bees feeding at the three distances, from closest distance, were respectively, 65.9 (1.29), 66.9 (1.65), and 67.0 (1.49) micrometer units. A one-way ANOVA performed on these data indicated heterogeneity in the sizes of bees visiting the feeder at the three distances (ANOVA; F = 6.77; $\underline{P}$ <.01; d.f. = 2,121). The bees sampled at the two longer distances were larger than those sampled 195 m from the hive. This experiment must be replicated, but the data from this and the previously described experiment suggest that the distance flow for food may be in part related to body size.

FLORAL CHOICE

Floral choices depend in part on the relationship between the morphology of pollinator and flower because the interaction between these factors affects foraging energetics. The foraging efficiency of bumble bees is influenced by proboscis length (Inouye 1980; Ranta and Lundberg 1980). Handling time (that is, time on a flower) (Inouye 1980) and rate of ingestion of nectar (Harder 1983) are both affected by proboscis length in Bombus. Morse (1978) captured bumble bees of one species foraging on the different size flowers of cow vetch (Vicia cracca L.), and found a positive relationship between corolla length and proboscis length. Similarly, the proboscis lengths of Bombus flavifrons Cresson workers foraging at the long-corolla larkspur (Delphinium) and the short-corolla bluebell (Mertensia) differed significantly.

Floral choice by honey bees is the topic of numerous investigations (see review by Waddington 1983a). Most investigations have been part of studies on the pollination biology of cultivated plants (Free 1970). The quality and quantity of nectar influences choice (Waller 1972), as do reinforcement schedule (Robacker and Ambrose 1981), flight distance between flowers (Marden and Waddington 1981), and handling time (Waddington 1985). Mackensen and Nye (1969) found preferences for alfalfa pollen based on genetic factors. However, I know of no published accounts of the relationship between floral choice and proboscis length or body size in honey bees.

Here I report one study that will facilitate beginning investigations of honey bee behavior as a function of proboscis length. The bee proboscis is difficult to measure because the glossa, which appears to be the functional unit of proboscis length, is flexible (Harder 1982; also see discussion in Morse 1978). Waddington and Herbst (1987) set out to estimate the "functional proboscis length" of honey bees. Operationally, functional proboscis length was the length of a capillary tube (diameter = 1 mm) emptied of sucrose solution by a feeding bee. It was a measure of the depth that the proboscis could be extended to reach nectar in a tubular corolla. Second, we examined whether measurements of other body parts could be used to predict functional proboscis length so that neither the proboscis length nor the functional proboscis length would need to be measured in the field. Waddington and

Herbst (1987) found that measurements on the wing
and head width were reasonable predictors of
functional proboscis length (Figure 13.6). The
wing measurements are particularly useful because
they can be performed on live, anesthetized bees
(Cideciyan 1984). It is left to be seen if floral
choice in honey bees is influenced by body size
and proboscis length.

VARIANCE IN BODY SIZE AND INTERACTIONS AMONG BEES

Thus far I have described some correlations
between some morphological measurements of
individuals and behavior of individuals.
Underlying the search for particular behavior
patterns to study and explanations of particular
correlations between behavior and body size is the
idea that certain tasks are performed best or most
efficiently by individuals having certain
morphological characteristics. A colony is a
collection of individuals that must complete a
number of tasks in order to survive and
reproduce. The success of the colony in passing
on genes, therefore, depends on the performance of
individuals in completing tasks. However, colony
fitness also depends on performance of groups of
individuals through communication and
interaction. Oster and Wilson (1978) used the
concept of "allometric space" to illustrate the
individual and social coverage of tasks and
ultimately to examine the optimal distribution of
castes in the colony in relation to tasks
requiring completion.
Each axis of the allometric space is a
measurement of the body, and together the
measurements provide the size and proportion of
individual workers (Figure 13.7). A task such as
foraging on a particular species of flower or
guarding may be performed best by a worker with
particular body size proportions. This perfect
worker can be indicated by a point in the
allometric space. This worker can also perform,
with moderate efficiency, tasks which are close by
in the allometric space (individual coverage).
Greater coverage is possible through social
interaction (social coverage). Based on this

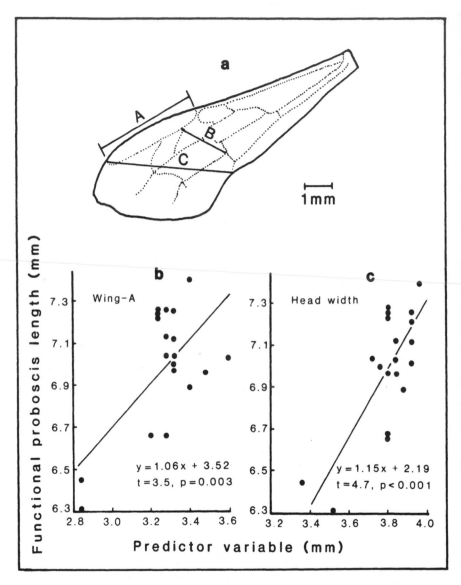

Figure 13.6. Simple linear regression analyses (Waddington and Herbst, unpublished data). (a) Three wing measurements were performed on bees. (b) Relationship between functional proboscis length and wing measurement A. Relationship between functional proboscis length and wing measurements B an C was similar to those with A. (c) Relationship between functional proboscis length and width of the head. N = 19 bivariate pairs.

construct Oster and Wilson (1978) modelled the mix
of castes that would yield the greatest
reproductive success to the colony given a set of
tasks. One prediction of their theoretical work
was that the colony should contain as many castes
as there are tasks to be performed. But, they
note, the number of tasks performed by social
insects always greatly exceeds the number of
distinct physical castes. They account for this
discrepancy in a number of ways, including the
high energetic cost of producing large forms,
behavioral plasticity of individuals, and the
limitations imposed by holometabolous
development. Social interactions may also account
for the low number of castes. For example, ants
may work together to haul large prey to the nest
or to ward off a predator.

Honey bees certainly perform an array of
tasks but have just a single worker physical
caste. Why is this? An obvious explanation is
the behavioral flexibility of workers. During its
life a single worker can be found constructing
comb, feeding brood, cleaning the nest, guarding,
and foraging at a variety of flowers, to name a
few jobs. Age polyethism may not be the only
answer.

Worker bees of almost all species are
monomorphic, but the amount of within-colony
variation in worker size varies considerably among
taxa. And, as indicated in the previous sections
of this chapter, this variation affects
behavior. Bumble bees and halictids have
considerable within-colony size variation, but the
honey bees and stingless bees have little. To
account for this difference Waddington (1981)
proposed an idea based on fundamental differences
in the foraging strategies of these taxa.

The most striking difference in the foraging
strategies between the closely related bumble bees
and honey bees or between halictids and stingless
bees is that bumble bees and halictids forage as
individuals only, but honey bees and stingless
bees recruit through a communication dance.
Bumble bees leave to sample flowers in the
vicinity of the colony. They appear to assess the
profitability of different floral species and
"major" on the most profitable species while

occasionally visiting one or more profitable
"minor" species (Heinrich 1976, 1979). Each
individual samples and then decides. A bee's
foraging returns depend on a number of
environmental factors including distance between
hive and flower patch, density of flowers, and
quality and quantity of nectar per flower
(Waddington 1983b). Furthermore, foraging
efficiency or profitability depends in part on

Figure 13.7. The concept of allometric space.
(a) Each microhabitat offers social insects a
potential task array. Each task can be most
efficiently dealt with by individual colony
members of a given body form $\underline{x}$, $\underline{y}$. A species
adapts by evolving a system of physical
castes that can be specified with a tightly
correlated allometric array, shown here as a
heavy curve. Efficiency contours have been
drawn around one point on the allometry
curve, which represents a single caste; each
caste performs tasks with declining
efficiency the farther they are from it on
the allometric space. Some task points are
close enough to the allometric curve to be
dealt with by individual colony members
(individual coverage); others can be
accommodated only through cooperative efforts
(social coverage). Still others cannot be
met at all; by definition these lie outside
the niche of the species. (b) The efficiency
function, $\underline{n}$, of a single caste $\underline{x}$, $\underline{y}$ is
projected over the allometric space. The
effectiveness of an average individual type
$\underline{x}$, $\underline{y}$ in the performance of a given task falls
off with the distance separating it from the
caste best suited to perform the task.
Beyond a certain distance, $\underline{\varepsilon}$, individuals
cannot perform the task with the minimum
level of efficiency required to make the
effort energetically profitable. (Figures
13.7a and b and legends are from G.F. Oster
and E.O. Wilson, 1978. Caste and ecology in
the social insects. Princeton New Jersey:
Princeton University Press, pp. 166, 183;
used with permission.)

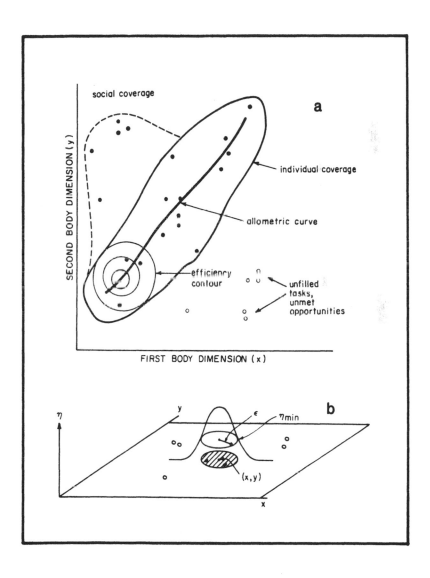

body size and especially proboscis length (Inouye 1980; Ranta and Lundberg 1980). Both the time it takes to visit a flower (handling time) and the amount of nectar reached inside the flower depend on proboscis length. Morse (1978) demonstrated that flowers are chosen on the basis of the relationship between corolla length and proboscis length. Variation in bumble bee worker size may be adaptive with respect to foraging because each bee forages on the flowers it perceives as "best" and bees of different size will utilize different flowers. High variance in body size results in the colony's efficient exploitation of a wider array of flora than if size variance were low because the effective density of usable flowers is higher and foraging costs are thereby reduced.

Scout honey bees choose among flowers as do the bumble bees; we know that honey bees also assess costs and intakes in order to decide among flowers (Waddington 1985). Recruits, which comprise the major proportion of a colony's foragers (Seeley 1983), are guided to flower patches by following dances. Recruits receive information on distance and direction to flowers and on profitability of flowers. Communication of profitability is not as well understood as other aspects of the recruitment dance; however, the scouts presumably integrate information on costs and intakes and send the message (Frisch 1967). Communicated information is based on the scout's perception of profitability, which is a result of its own foraging experience.

I previously pointed out that if foraging costs and intakes experienced by honey bees are affected by body size, as they are in bumble bees, then information sent on profitability is accurate in terms of what the recruit will experience in the field only if the scout and recruit are the same size (Figure 13.8) (Waddington 1981). A larger bee with a longer proboscis may reach to the bottom of corollas of a particular plant species and find it profitable to forage. This bee would communicate "high profitability" and this message would be received by all bees, including smaller bees that cannot reach so far into the corollas and would not find this species profitable to forage. Such a transfer of

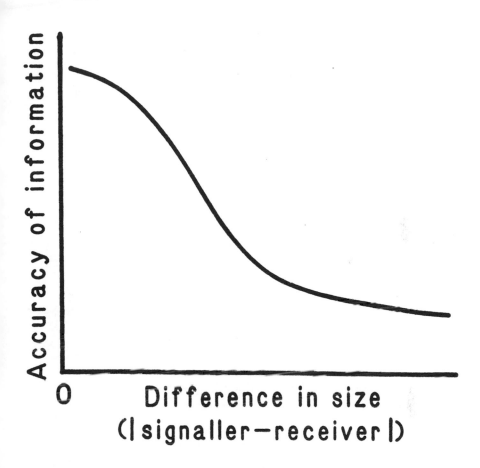

Figure 13.8. Hypothetical relationship between
accuracy of information sent on flower
profitability (in terms of recruit's
experience) and the absolute difference
between the scout's and recruit's body size.

misinformation would be costly since bees would be
guided to flowers on which they do not forage
maximally. The amount of misinformation sent in
the colony is increased by increased variance in
body size of workers. Since transfer of
misinformation is costly, it follows that the rate
of food return by the colony is inversely related
to size variance. Using the idea of allometric
space (Oster and Wilson 1978), the colony could
exploit more flowers above some minimum efficiency
with increased size variance (Figure 13.9).

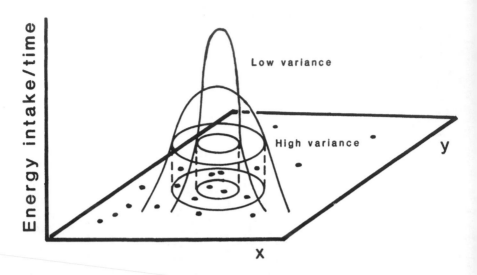

Figure 13.9. Honey bees forage at some flower
    species in this allometric space. Having
    high worker size variance permits the colony
    to forage more species at a profit than if
    variance is low. However, the colony's
    profits are greater when variance is low
    because the recruitment system is more
    effective than when variance is high.

However, when within-colony size variance is low
the colony's efficiency of foraging is increased
by the enhanced effectiveness of the recruitment
system, even though fewer flowers can be foraged
with minimum efficiency. The recruitment system
permits the colony to shift its foraging effort
quickly to the few most rewarding flower patches
(Visscher and Seeley 1982). This expected inverse
relationship between size variance and colony
feeding efficiency may explain the general pattern
of within-colony worker body size variance among
species. Workers of the highly eusocial honey
bees and stingless bees, which have recruitment
systems, have lower variance in body size than
bees without recruitment systems (Michener 1974;
Waddington 1981).
    I have twice tested predictions of the size
variance-communication hypothesis; in each case
the results were consistent with prediction.
These studies are summarized below.

Stingless bees (Melponini) have, in general, little within-colony variation in worker body size. I looked at variation in size among species of stingless bees because complexity of recruitment systems varies greatly among species, from primitive systems to systems as derived as that of honey bees. For example, Trigona silvestrii Friese scouts notify other workers of food, but no information is communicated on the spatial position of flowers in relation to the hive. This is just a small (but important) step above the situation in bumble bees. In contrast, some Melipona species communicate distance and direction to flowers. I predicted that species with more derived recruitment systems would have lower size variance than those species with rudimentary recruitment systems (Waddington et al. 1986).

Workers from 14 nests comprising 11 species of stingless bees were collected in Costa Rica and Panama. Size was quantified using various measures of sclerotized body parts and size variation (coefficient of variation) was quantified for each colony. Species were ranked according to step in the evolution of recruitment system (Kerr 1969) (as well as "evolutionary level" [Wille 1979] and effectiveness of recruitment). Results of Kendall correlation analysis indicated a highly significant negative relationship between size variance and complexity of recruitment system (-0.74, P <.001). This inverse relationship among the stingless bees mirrors the general pattern among all bees. As a cautionary note, this correlative study suffers from the same weakness of almost any correlative study: the degree of causation is unknown. For example, in this study phylogenetic rank and size variance were also significantly correlated, as were phylogenetic rank and complexity of the recruitment system. What affects what?

A more direct test of the effects of size variation on the colony was undertaken. Waddington (1981) proposed that the effectiveness of the recruitment system would be adversely affected by variance in body size. Cost to the colony would be in the form of reduced rate of food return with increased size variance. I tested the prediction of an inverse relationship between colony size variance and rate of food intake in the honey bee. Variance in body size of foragers was periodically sampled in six colonies over six weeks. When size variance was sampled so was the colony's rate of nectar intake. Nectar intake was calculated as the difference in weight between samples of bees returning and leaving the colony, and the rate that bees returned to each colony during each sampling period was determined. Other factors that might affect rate of nectar intake were also sampled: population size, mean body size of foragers, sampling date, and sampling time.

Partial correlation analyses were used to test the relationship between variance in body size and the rate of nectar intake. I controlled for variance in the above four factors that might also affect rate of intake. Rate of nectar intake was negatively correlated with size variance in each colony, but the partial correlation coefficients were significantly different from zero in two of the six colonies (Table 13.4). The probabilities from the six partial correlation analyses may be combined to detect a general negative relationship (Sokal and Rohlf 1969:623). Results of this procedure were statistically significant (X^2 = 25.9, d.f. = 10, $\underline{P}$ <.025).

These results are consistent with predictions of the size variance-communication hypothesis; however, the results do not strongly support the hypothesis. This may be because bees that make up dance groups (a scout and the dance followers) are similar in size; that is, variance in body size within dance groups is lower than the colony's overall variance (Waddington 1981). In this way bees could at least partially obviate the possible negative effects of size variance because communication occurs between bees of similar size. I sampled on two occasions in the same colony scout (dancer) and dance followers of several dance groups and measured the bees (measurement D, Figure 13.3) (Waddington, unpublished data). Results of one-way ANOVAs performed on each of the two data sets with dance groups being groups of the ANOVA showed that

Table 13.4
Partial correlations between variance in body size and the colony's rate of nectar intake (mg/min)

| | \multicolumn{6}{c}{Colony} | | | | |
	1	2	3	4	5	6
Coefficient	-0.45	-0.35	-0.46	-0.41	-0.01	-0.23
Probability	0.08	0.44	0.02	0.04	0.49	0.17

variance within groups was indeed lower than
between groups. Size of the dancer was also
significantly (positively) correlated with size of
dance followers. It is not known how these dance
groups of similarly sized bees form, but the fact
certainly could play a role in colony recruitment
efficiency.

CONCLUSION

Little is known about how size affects
behavior and task performance of worker honey
bees. Therefore, the purpose of this chapter was
not to review a vast literature and present a
nearly final story. Rather, preliminary data and
weakly tested ideas are presented in order to
stimulate research on potentially interesting
questions. At this point I conclude from the few
available data that body size (or perhaps unknown
correlates of body size) affects behavior and that
discovered and undiscovered effects may play a
role in colony ergonomics. If workers are found
to distribute tasks in part according to size,
implying adaptive behavior, the next step will be
to investigate the colony's worker size
distribution in relation to variation in the
distribution of tasks; can a colony manipulate its
body size distribution in a way that approaches
maximization of completion of a particular set of
tasks? Studies of body size-behavior
relationships should be assessed on the basis of
both individual worker performance and performance
in the social context.

SUMMARY

The effects of body size on behavior of
individual honey bees (Apis mellifera L.) and on
social interactions between bees are explored.
Age of first foraging, frequency of foraging, and
distance flown from the hive to flowers are
affected by body size. Relative body size of a
scout and recruits involved in recruitment is
hypothesized to inversely affect accurate
transmission of information on foraging

profitability. This idea yields several predictions which are tested. (1) There is an inverse relationship between within-nest variation in body size of species of stingless bees and the complexity of the species' recruitment system. (2) Size variation of members within honey bee dance groups (a dancing scout and recruits) is lower than size variation among dance groups. (3) Rate of a colony's intake of nectar is sometimes inversely related to the variation in size of foragers. Work on relationships between body size and behavior in bees is in a very early state of investigation.

ACKNOWLEDGMENTS

My research reported in this chapter was supported by NSF grant #DEB 8119280 and by the American Philosophical Society. Larry Herbst and Charles D. Michener made useful comments on a draft of the manuscript. Burger King on Bird Road provided space and coffee for writing.

REFERENCES

Abdellatif, M.A. 1965. Comb cell size and its effects on the body weight of the worker bee, Apis mellifera L. American Bee Journal 105:86-87.

Boch, R. 1956. Die Tanze der Bienen bei nahen und fernen Trachtquellen. Zeitschrift für Vergleichende Physiologie 38:136-167.

Cideciyan, M. 1984. The relationship between size and behavior in worker honey bees (Apis mellifera). Master's thesis, University of Miami, Coral Gables.

Free, J.B. 1965. The allocation of duties among worker honey bees. Zoological Society of London, Symposium 14:39-59.

Free, J.B. 1970. Insect pollination of crops. New York: Academic Press.

Frisch, K. von 1967. The dance language and orientation of bees. Cambridge,

414

Massachusetts: Harvard University Press.
Gary, N.E. 1975. Activities and behavior of
honey bees. In The hive and the honey bee,
ed. Dadant and Sons, 185-264. Hamilton,
Illinois: Dadant and Sons.
Harder, L.D. 1982. Measurement and estimation of
functional proboscis length in bumble bees
(Hymenoptera: Apidae). Canadian Journal of
Zoology 60:1073-1079.
Harder, L.D. 1983. Flower handling efficiency of
bumble bees: morphological aspects of
probing time. Oecologia 57:274-280.
Heinrich, B. 1976. The foraging specializations
of individual bumble bees. Ecological
Monographs 46:105-128.
Heinrich, B. 1979. "Majoring" and "minoring" by
foraging bumble bees, Bombus vagans: an
experimental analysis. Ecology 60:245-255.
Inouye, D. 1980. The effect of proboscis and
corolla tube length on patterns and rates of
flower visitation by bumble bees. Oecologia
45:197-201.
Kerr, W.E. 1969. Some aspects of the evolution
of social bees. Evolutionary Biology 3:119-
175.
Kerr, W.E., and Hebling, N.J. 1964. Influence of
the weight of worker bees on division of
labor. Evolution 18:267-270.
Kulzhinskaya, K.P. 1956. The role of the food
factor in the growth of bees. Apicultural
Abstracts 37:177.
Lindauer, M. 1952. Ein Beitrag zur Frage der
Arbeitsteilung im Bienenstaat. Zeitschrift
für Vergleichende Physiologie 34:299-345.
Lindauer, M. 1961. Communication among social
bees. Cambridge, Massachusetts: Harvard
University Press.
Mackenson, O., and Nye, W.P. 1969. Selective
breeding of honeybees for alfalfa pollen
collection: sixth generation and
outcrosses. Journal of Apicultural Research
8:9-12.
Marden, J.H., and Waddington, K.D. 1981. Floral
choices by honeybees in relation to the
relative distances to flowers. Physiological
Entomology 6:431-435.
Michener, C.D. 1974. The social behavior of the

bees. Cambridge, Massachusetts: Harvard University Press.

Morse, D.H. 1978. Size related foraging differences of bumble bee workers. Ecological Entomology 3:189-192.

Oster, G.F., and Wilson, E.O. 1978. Caste and ecology in the social insects. Princeton, New Jersey: Princeton University Press.

Otto, D. 1958. Über die Arbeitsteilung im Staate von Formica rufa rufo-pratensis minor Gössw. und ihre verhaltensphysiologischen Grundlagen: Ein Beitrag zur Biologie der roten Waldameise. Wissenschaftliche Abhandlungen der Deutschen Akademie der Landwirtschaftswissenschaften, Berlin 30:1-169.

Plowright, R.C., and Jay, S.C. 1977. On the size determination of bumble bee castes (Hymenoptera: Apidae). Canadian Journal of Zoology 55:1133-1138.

Ranta, E., and Lundberg, H. 1980. Resource partitioning in bumble bees: the significance of differences in proboscis length. Oikos 35:298-302.

Ribbands, C.R. 1952. Division of labour in the honeybee community. Proceedings of the Royal Society (B) 140:32-43.

Ribbands, C.R. 1964. The behaviour and social life of honeybees. New York: Dover Publications, Inc.

Robacker, D.C., and Ambrose, J.T. 1981. Effects of partial reinforcement on recruiting behaviour in honeybees foraging near the hive. Journal of Apicultural Research 20:19-22.

Roberts, W.C. 1956. Hybrid vigor in honeybees. Proceedings of the XVI International Beekeepers Conference (Apicultural Abstracts 246:56).

Ruttner, F., and Mackensen, O. 1952. The genetics of the honeybee. Bee World 33:53-63, 71-79.

Seeley, T.D. 1982. Adaptive significance of the age polyethism schedule in honeybee colonies. Behavioral Ecology and Sociobiology 11:287-293.

Seeley, T.D. 1983. Division of labor between

scouts and recruits in honeybee foraging. Behavioral Ecology and Sociobiology 12:253-259.

Sekiguchi, K., and Sakagami, S.F. 1966. Structure of foraging population and related problems in the honeybee, with considerations on the division of labour in bee colonies. Report of the Hokkaido National Agricultural Experiment Station 69:1-65.

Sokal, R.R., and Rohlf, F.J. 1969. Biometry. San Francisco: W.H. Freeman and Company.

Visscher, P.K., and Seeley, T.D. 1982. Foraging strategy of honeybee colonies in a temperate deciduous forest. Ecology 63:1790-1801.

Waddington, K.D. 1981. Patterns of size variation in bees and evolution of communication systems. Evolution 35:813-814.

Waddington, K.D. 1983a. Foraging behavior of pollinators. In Pollination biology, ed. L. Real, 213-239. New York: Academic Press.

Waddington, K.D. 1983b. Floral-visitation-sequences by bees: models and experiments. In Handbook of experimental pollination biology, ed. C.E. Jones, R.J. Little, 461-473. New York: Van Nostrand Reinhold Co. Inc.

Waddington, K.D. 1985. Cost-intake information used in foraging. Journal of Insect Physiology 31:891-897.

Waddington, K.D. and Herbst, L.H. 1987. Body size and the functional length of the proboscis of honey bees. The Florida Entomologist 70:124-128.

Waddington, K.D., Herbst, L.H., and Roubik, D.W. 1986. Relationship between recruitment systems of stingless bees and within-nest worker size variation. Journal of the Kansas Entomological Society 59:95-102.

Waller, G.D. 1972. Evaluating responses of honeybees to sugar solution using an artificial-flower feeder. Annals of the Entomological Society of America 65:857-862.

Wells, P.H., and Giacchino, J. 1968. Relationships between the volume and the sugar concentration of loads carried by honeybees. Journal of Apicultural Research 7:77-82.

Wille, A. 1979. Phylogeny and relationships among the genera and subgenera of the stingless bees (Meliponinae) of the world. Revista de Biologia Tropical 27:241-277.

Wilson, E.O. 1953. The origin and evolution of polymorphism in ants. Quarterly Review of Biology 28:136-156.

Wilson, E.O. 1971. The insect societies. Cambridge, Massachusetts: Harvard University Press.

Wilson, E.O. 1980. Caste and division of labor in leaf-cutter ants (Hymenoptera: Formicidae: Atta). I. The overall pattern in A. sexdens. Behavioral Ecology and Sociobiology 7:143-156.

14

Elitism in Social Insects:
A Positive Feedback Model

R. C. Plowright and C.M.S. Plowright

If one spends time watching a bumble bee
colony in which the workers are given
distinguishing marks so that they can be
recognized individually, it usually turns out that
some individuals work much harder than others.
This is true whether the activity is collecting
food for the colony or carrying out duties within
the hive, such as feeding larvae. Usually, for
example, some individuals repeatedly return with
huge pollen loads in their corbiculae after short
foraging trips of half an hour or less. Others,
flying from the same colony on the same day, may
be seen returning with almost no nectar or pollen
after an absence of several hours. Similarly, in
Bruce Pendrel's work (1977) on larval feeding, a
wide spread in the level of working activity among
colony members within the hive was consistently
found: some would distinguish themselves by a
steady level of high job performance while others
did almost no work for hours at a time.
  Such variability in activity level among
adult worker individuals is well known among other
social insects. For example, Chen (1937a, b)
documented hyperactivity in nestbuilding for some
individuals of Camponotus japonicus var.
aterrimus Emery. Lenoir (1981) reported that four
percent of individuals in a colony of Lasius niger
L. carried out more than ten percent of all
transports in a brood-retrieving test. Indeed,
the phenomenon of work variability is so general
that it has prompted explanations regarding its
functional significance. One suggestion, that the

idle members of the work force constitute a
reserve that can be called upon during times of
stress (Lindauer 1961), is particularly appealing
because it neatly reconciles the phenomenon with
evolutionary principles.

However, our present concern is not with the
functional significance of activity variation but
rather with the proximate causal mechanisms which
bring it into being. In particular, we address
the question of how proximate these mechanisms
really are. For purposes of discussion, let us
suppose that the workers in a bumble bee colony
can be divided, more or less cleanly, into two
classes: elites (who do almost all the work) and
loafers (who do very little). Such a crisp
categorization may not apply very well in any real
case, but for the moment we use it to achieve a
clear exposition. The causal question then
becomes, How do the elites become so hardworking,
and how do the loafers acquire their idle habits?

One may recognize two extremes along the
spectrum of possible answers to this question. On
the one hand, it might be that loafers and elites
are differentiated genetically. On the other,
perhaps they are merely the chance products of
simple probabilistic processes, requiring little
more by way of explanation than a population of
coins, some of which have landed on their heads
and the rest on their tails. This paper is
concerned with the second of these alternatives,
although it must immediately be remarked that a
purely stochastic explanation, at least in its
extreme form, seems to be ruled out by the
facts. For example, Pendrel and Plowright (1981)
have shown that the activity of bumble bee workers
can be modified by changing the level of job
opportunities in the colony. The presence of more
hungry larvae appears to stimulate a higher level
of larval feeding activity, and so on. Similarly,
increasing the number of nymphs to be transported
in a colony of Tapinoma erraticum Latr. led not
only to a recruitment of workers to transport
them, but also to an increase in the activity
level of previously transporting ants (Meudec
1977). Nevertheless, since what we are concerned
with is variance in work activity, rather than its
general level, the possibility that probabilistic

processes are in operation is not entirely excluded.

A precedent for this possibility is to be found in the work of Oster (1976), who proposed a semi-Markovian structure to account for the genesis of division of labor in bumble bees (and, by implication, in other social insects). We were struck by similarities between Oster's model and some other models of motivation in the ethological literature, e.g. Heiligenberg's (1974) account of processes governing behavioral states of readiness in crickets. Accordingly, we decided to run Monte Carlo simulations of a semi-Markovian description of job performance in a bee hive, hoping that the results might throw light on the origin of differences between elites and loafers.

Figure 14.1 illustrates the basic structure of the model. The activity level of each bee with respect to each of the range of possible job opportunities is governed by both external and internal factors. The external factors consist merely of the present levels of 'employment opportunities' within the colony: the probability that a bee will decide to feed larvae is increased by the presence of larvae that are hungry, and foraging is stimulated by the presence of empty honey or pollen pots. The internal factors, on the other hand, may be thought of as something analogous to 'job satisfaction.' We term them positive feedback factors to capture the essence of Oster's suggestion that insects which do a job successfully are thereby more likely to repeat that particular type of activity when the opportunity is next presented. Such a suggestion has also been made, more recently, by Lenoir and Ataya (1983). As can be seen in Figure 14.1, the positive feedback factor is increased by job performance but attenuates when that particular type of job has not recently been carried out.

The flow chart shown in Figure 14.2 defines the model more precisely. The real business of the simulation is captured by the statement

$$P = 1 - e^{-IKE},$$

422

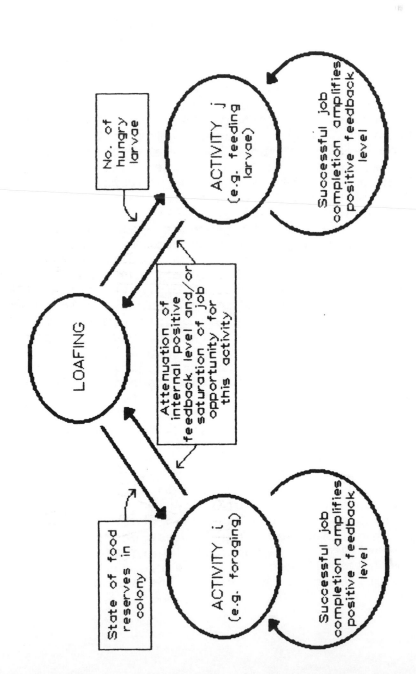

Figure 14.1. Basic structure of positive feedback model of elitism.

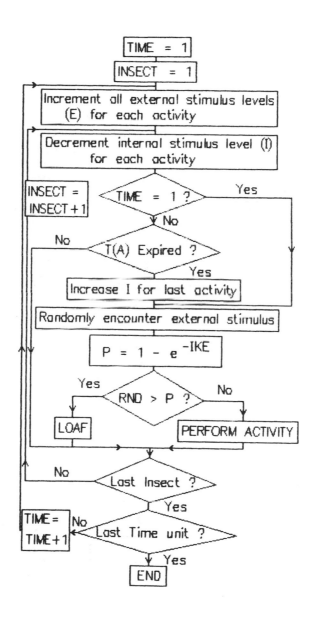

Figure 14.2.   Flow chart of simulation of events
within a social insect colony.

where P is the probability that, once having encountered a task of some specified type, the bees will actually perform that task; K is a scaling constant; and I and E are, respectively, the levels of the internal and external stimuli.

We ran the model for a colony containing 100 worker insects and over a period of 1000 time units (following an initial period of 500 time units for the system to settle down). By adjusting the parameter values for the rate of decay and minimum possible level of the positive feedback factor, we obtained three broad classes of result, shown in Figure 14.3. When the positive feedback factor was allowed to decay rapidly during a period when no job was being performed, the bees became unimodally idle (Figure 14.3A). (Figure 14.3A is very similar to one given by Abraham and Pasteels [1980] showing the variation of individual activity in nest-building by Myrmica rubra L.) When, by contrast, this factor was set to decay only slowly and was constrained not to fall below a relatively high minimum level, we succeeded in transforming our insects into workaholics (Figure 14.3B). The third result, obtained by setting a very low minimum possible level for the positive feedback factor, is perhaps the most interesting: a bimodal distribution of bees is generated, with a differentiated class of loafers on the one hand, and a group of elites on the other (Figure 14.3C).

Bimodal frequency distributions such as this are well-known outcomes of certain types of stochastic models. A recent biogeographical example is provided by Hanski's (1982) "Core-Satellite Hypothesis" of regional distribution. Typically, in these cases the bimodal result is the result only of the stochastic version of the model; the deterministic version, by contrast, gives only what corresponds to the mean value of the bimodal distribution.

Demonstrating that it is possible to obtain a bimodal outcome from a computer model is not in itself, of course, evidence that the phenomenon of elitism in social insects is merely the ephemeral outcome of a set of stochastic processes. However, we hope that we have demonstrated that

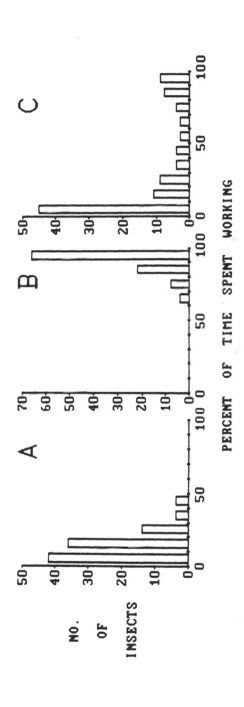

Figure 14.3. Frequency of individuals at various levels of activity (summed over two tasks) generated by Monte Carlo simulations. (A) distribution skewed to the right: fast decay of positive feedback factor. (B) distribution skewed to the left: high minimum positive feedback factor. (C) bimodality: low minimum positive feedback factor.

such stochastic processes could, in principle, account for the observed results.

Our model also mimics real bee behavior in one other significant respect. Several years ago, when working with the fierce Brazilian bumble bee, Bombus atratus Fkln., we observed (Plowright and Roach, unpublished data) an 'overshoot response' in foraging activity after periods of food deprivation imposed on the colony. When food was again provided in the flight room in which the bees were allowed to forage, for awhile they foraged at a much greater rate than normal. This was followed by a second phase in which foraging activity then fell below the normal rate, perhaps because the honey pots in the colony had become abnormally full. So, the overall pattern of resumption of foraging activity had the form of a series of oscillations that gradually became damped until the insects were once again foraging at their average rate.

We set up analogous experimental conditions in our computer model. As shown in Figure 14.4, we simulated an experimental intervention by removing all food stores from the colony. In comparison with unmanipulated control runs, the experimental run gave results (Figure 14.4) that are qualitatively in agreement with the B. atratus observations: foraging activity, in the time period following the removal of food, showed the characteristic overshoot tendency. While, again, this is not in itself evidence that the model is in any way an accurate representation of the workings of bee behavior, we were encouraged that another of its outcomes is consistent with empirical findings.

Although, as Oster and Wilson (1978) point out, most reports of elitism pertain to levels of activity for only a single task, cases of a more general elitism have also been recorded: they cite the work of Otto (1958), in which some individual ants were found to be more active than others with respect to several tasks. While typically a marked division of labor was obtained in our simulations, in the sense that individuals active with respect to one task tended to be inactive with respect to the other, we occasionally found individuals who worked hard at

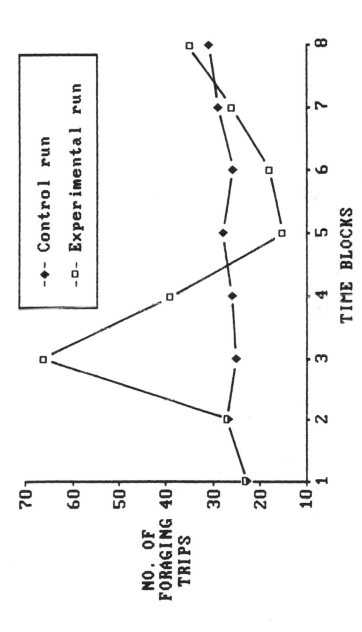

Figure 14.4. Overshoot effect generated by model of elitism. Each block consists of sixty time units. In the experimental run all food was removed from the colony at the beginning of the third time block.

both tasks. Further work would be needed,
however, to decide whether our model is capable of
accounting for the origin of generalized elitism.

In conclusion, we would like to step back and
offer some general comments about our attempt to
explain elitism in social insects. In line with a
contemporary trend that demands the use of
explicit null-models in ecology and ethology, we
support the development of explanations of insect
social behavior that begin with a minimum amount
of assumed structure. The simple model that we
have presented here is offered in this spirit.
The difficulty that we have not yet quite resolved
in our own minds is how one proceeds when such a
'null-model' is found to be acceptable. In the
present case, a model that contains only a few
parameters can, as we have reported, readily be
tuned to mimic a range of real-world phenomena.
Yet, as we have mentioned repeatedly, merely to
show that a simple stochastic model is capable of
yielding a facsimile of the elitism phenomenon
falls far short of demonstrating that elitism
results only from simple stochastic processes. To
go further, it seems, will necessitate much more
careful ethological studies aimed at estimating
the relevant parameters from real colonies of real
social insects.

In order to add yet more caution to the
interpretation of our results, we make what
amounts to a paraphrase of thermodynamic
principles: if our model (or presumably any other
similar stochastic model of activity level and
division of labor in a bee hive) is run long
enough, the effects of the initial conditions are
gradually dissipated. By this we mean that there
occurs a gradual kaleidoscopic shifting of task
specializations among the colony members. But
when one reads the French ant literature (as far
as we have been able to determine, the term
'elite' was first applied to social insects by
Combes in 1937), for example Meudec (1973), one is
struck by the persistence of the quality called
elitism. Meudec reported that 10 percent of those
workers that were most active on the first day of
observations remained so on each of the subsequent
days of her 10-day observation period. Similarly,
over a period of six weeks, in a colony of

Tapinoma erraticum subjected to nest disturbance tests, one worker was almost always inactive while another was nearly always carrying brood (Meudec and Lenoir 1982). Stability of high activity levels has also been reported by Verron (1976) for *Lasius niger.* These results perhaps show the inadequacy of simple stochastic explanations, such as those built into our computer model, to account for the development of elitism over a larger time frame (see Pendrel and Plowright [1981] for a discussion of persistence of task specificity over different time scales with respect to Oster's model).

SUMMARY

A simple stochastic model, incorporating a semi-Markovian transition structure based on positive feedback processes (cf. Oster 1976), was written to simulate task performance and division of labor in a social insect colony. The model could be tuned to generate some aspects of the phenomenon of elitism in insect societies. The authors discuss the relevance of models of this sort with respect to inferences about causation in insect social behavior.

ACKNOWLEDGMENTS

We gratefully record our debt to Professor E.O. Wilson for repeatedly stimulating our interest in the phenomenon of elitism in social insects. We thank Stefan Cover and Jerry Hogan for helpful discussions and for providing us with several pertinent references.

REFERENCES

Abraham, M., and Pasteels, J.M. 1980. Social behaviour during nest-moving in the ant *Myrmica rubra* L. (Hym. Form.). *Insectes Sociaux* 27:127-47.

Chen, S.C. 1937a. Social modification of the activity of ants in nest-building. Physiological Zoology 10:420-36.

Chen, S.C. 1937b. The leaders and followers among the ants in nest-building. Physiological Zoology 10:437-55.

Combes, M. 1937. Existence probable d'une élite non différenciée d'aspect, constituant les véritables ouvrières chez les Formica. Comptes Rendus de l'Académie des Sciences, Paris 204:1674-75.

Hanski, I. 1982. Dynamics of regional distribution: the core and satellite species hypothesis. Oikos 38:210-221.

Heiligenberg, W. 1974. Processes governing behavioral states of readiness. In Advances in the study of behavior, ed. D.S. Lehrman, J.S. Rosenblatt, R.A. Hinde, and E. Shaw, vol. 5, 173-200. New York: Academic Press.

Lenoir, A. 1981. Brood retrieving in the ant, Lasius niger L. Sociobiology 6:153-78.

Lenoir, A., and Ataya, H. 1983. Polyéthisme et répartition des niveaux d'activité chez la fourmi Lasius niger L. Zeitschrift für Tierpsychologie. 63:213-232.

Lindauer, M. 1961. Communication among social bees. Cambridge, Massachusetts: Harvard University Press.

Meudec, M. 1973. Sur les variations temporelles du comportement de transport du couvain dans un lot d'ouvrières de Tapinoma erraticum Latr. (Formicidae Dolichoderinae). Comptes Rendus de l'Académie de Science, Paris, Series D 277:437-440.

Meudec, M. 1977. Le comportement de transport du couvain lors d'une perturbation du nid chez Tapinoma erraticum (Dolichoderinae). Rôle de l'individu. Insectes Sociaux 24:345-52.

Meudec, M., and Lenoir, A. 1982. Social responses to variation in food supply and nest suitability in ants (Tapinoma erraticum). Animal Behaviour 30:284-92.

Oster, G.F. 1976. Modeling social insect populations. I. Ergonomics of foraging and population growth in bumblebees. American Naturalist 110:215-245.

Oster, G.F., and Wilson, E.O. 1978. Caste and ecology in the social insects. Princeton, New Jersey: Princeton University Press.

Otto, D. 1958. Über die Arbeitsteilung im Staate von Formica rufa rufo-pratensis minor Gössw. und ihre verhaltensphysiologischen Grundlagen: Ein Beitrag zur Biologie der roten Waldameise. Wissenschaftliche Abhandlungen der Deutschen Akademie der Landwirtschaftswissenschaften, Berlin 30:1-169.

Pendrel, B.A. 1977. The regulation of pollen collection and distribution in bumble bee colonies (Bombus Latr.: Hymenoptera). M.Sc. thesis, Department of Zoology, University of Toronto.

Pendrel, B.A., and Plowright, R.C. 1981. Larval feeding by adult bumble bee workers (Hymenoptera: Apidae). Behavioral Ecology and Sociobiology 8:71-76.

Verron, H. 1976. Note sur la stabilité de certains traits éthologiques chez les ouvrières de Lasius niger (Hyménoptère Formicidae). Comptes Rendus de l'Académie de Science, Paris, Series D 283:671-74.

About the Contributors

PRASSEDE CALABI recently completed a Ph.D. in
demography of caste in ants at Boston
University. She is currently working with the
graduate program in extension at Harvard
University. Her research interests include
division of labor, caste and ecology of social
insects. [Graduate Program in Extension, Harvard
University, 20 Garden St., Cambridge,
Massachusetts, 02138].

NORMAN F. CARLIN received his Ph.D. in biology in
1986 at Harvard University, where he is currently
a postdoctoral research associate. He is
principally interested in recognition behavior and
the role of queens in ant colonies. [Museum of
Comparative Zoology, Harvard University,
Cambridge, Massachusetts, 02138].

BLAINE COLE completed the Ph.D. in biology at
Princeton University in 1979, then held
postdoctoral positions at Harvard University, the
University of Utah and the University of
California, Berkeley. He is currently an
assistant professor in the biology department at
the University of Virginia. He is also the
director of the Mountain Lake Biological Station,
where he spends his summers. His major interests
are in the evolution of sociality, the
organization of colony functioning, the relation
between individual-level and group-level
selection, and slave making in the genus
Formica. [Department of Biology and Mountain Lake

434

Biological Station, University of Virginia,
Charlottesville, Virginia, 22903].

HOLLY A. DOWNING received her BA in zoology at
Smith College in 1978 and her Ph.D. in entomology
at the University of Wisconsin, Madison in 1986.
Her research interests are insect social behavior
and nest construction. [Biology Department,
University of Wisconsin-Whitewater, Whitewater,
Wisconsin, 53190].

ERIC H. ERICKSON, JR., is supervisory research
entomologist and research leader at the Carl
Hayden Bee Research Center, Tucson, Arizona. He
received his Ph.D. in entomology from the
University of Arizona. His principal research
interests involve basic and applied research into
the behavior and biology of honey bees and related
Hymenoptera. [U.S. Department of Agriculture,
Agricultural Research Service, Carl Hayden Bee
Research Center, Tucson, Arizona, 85719].

JENNIFER H. FEWELL is working on her Ph.D. at the
University of Colorado. She received her B.A.
from Cornell University and her M.A. at the
University of Colorado. Her major research
interest is the role of flexibility in the
foraging systems of social insects. [Department
of EPO Biology, University of Colorado, Boulder,
Colorado, 80309].

DOMINIQUE FRESNEAU obtained the degree of Docteur
de troisième cycle from the Université Paris V in
1976 and has been Maitre de Conférences at the
Université Paris Nord since 1978. He is presently
working on a research program concerning the
social organization in primitive ant species.
[Laboratoire d'Ethologie et Sociobiologie,
Université Paris XIII, Av. J.-B. Clément, F-93430,
Villetaneuse, France].

DEBORAH M. GORDON received the B.A. in French from
Oberlin College, the M.S. in Biology from Stanford
University, and the Ph.D. in Zoology from Duke
University,. She then was a junior fellow of the
Harvard Society of Fellows before moving to
Oxford. [Centre for Mathematical Biology,

Mathematics Institute, University of Oxford, 24-29
St. Giles, Oxford OX1 3LB, England].

PIERRE JAISSON obtained the degree of Docteur
d'Etat ès-Sciences from the Université Pierre et
Marie Curie in 1975. Since 1977 he has been
Titular Professor of Ethology at the Université
Paris-Nord, where he founded the Laboratoire
d'Ethologie et Sociobiologie. Professor Jaisson
is the Scientific Editor of the international
journal Insectes Sociaux, the organ of the
International Union for the Study of Social
Insects. He works on early learning involved in
recognition processes in ants. [Laboratoire
d'Ethologie et Sociobiologie, Université Paris
XIII, Av. J.-B. Clément, F-93430, Villetaneuse,
France].

ROBERT L. JEANNE received the Ph.D. in biology
from Harvard University in 1971, and is now
professor of entomology and zoology at the
University of Wisconsin. He studies
communication, nesting behavior, defense, and the
organization of work in social wasps, primarily in
the tropics. [Department of Entomology,
University of Wisconsin, Madison, Wisconsin,
53706].

JEAN-PAUL LACHAUD is investigator at the Centre
National de la Recherche Scientifique and obtained
the degree of Docteur de troisième cycle from the
Université Paul Sabatier (Toulouse). He is
presently in the Université Paris Nord, working on
a research program concerning the social behavior
of some Neotropical ant species. [Laboratoire
d'Ethologie et Sociobiologie, Université Paris
XIII, Av. J.-B. Clément, F-93430, Villetaneuse,
France].

CATHERINE PLOWRIGHT is working toward her Ph.D. in
psychology at the University of Toronto. She also
enjoys bumble bees (especially catching queens for
domestication in the spring), but her dissertation
research involves analyses of avian foraging
behavior. [Department of Psychology, University
of Toronto, Toronto, M5S 1A1, Ontario, Canada].

R. CHRISTOPHER PLOWRIGHT is an associate professor of zoology at University of Toronto. Now 50 years old, he has been captivated by an obsessive interest in bumble bees since his 14th year. He and his associates also work on aspects of pollination ecology. [Department of Zoology, University of Toronto, Toronto, M5S 1A1, Ontario, Canada].

GREGORY B. POLLOCK is a doctoral candidate at the School of Social Science, University of California, Irvine. He currently holds a MacArthur Foundation Fellowship in International Peace and Security granted by the Social Science Research Council. His research interests focus on the effects of population structure on the evolution of cooperation and conflict. [School of Social Science, University of California, Irvine, California, 92717].

DAVID C. POST received his Ph.D. in entomology at the University of Wisconsin, Madison, in 1984. His research interests are in the behavioral ecology of the paper wasp Polistes and of Apis mellifera. [Biology Department, University of Wisconsin-Whitewater, Whitewater, Wisconsin, 53190].

STEVEN W. RISSING is an assistant professor of zoology at Arizona State University. He received his Ph.D. in zoology from the University of Washington in 1980 and held a postdoctoral fellowship in behavioral biology at the University of Chicago 1980-1981. His research interests center on the behavioral ecology of social insects, especially desert seed-harvester ants. [Department of Zoology, Arizona State University, Tempe, Arizona, 85287]

REBECA B. ROSENGAUS received her B.A. in Biology at Boston University in 1984 and currently she is a M.A.-Ph.D. graduate student at Boston University. Her research interests include insect social behavior and its evolution, particularly in termites. [Department of Biology, Boston University, 2 Cummington St., Boston, Massachusetts, 02215].

JAMES F. A. TRANIELLO is assistant professor of biology at Boston University. After receiving his Ph.D. from Harvard University in 1980, he was appointed lecturer in the Department of Organismic and Evolutionary Biology at Harvard, and subsequently moved to Boston University in 1981. His research interests are in insect behavioral ecology, particularly in foraging behavior and caste evolution in ants and termites. [Department of Biology, Boston University, 2 Cummington St., Boston, Massachusetts, 02215].

ROBERT K. VANDER MEER received his Ph.D. in synthetic organic chemistry from the Pennsylvania State University, after which he served in the Peace Corps for four years as assistant professor of chemistry at the University of the South Pacific, Suva, Fiji Islands. He then spent a year and a half at Cornell University doing postdoctoral work on the chemical ecology of the moth, Utelheisa ornatrix, before moving to his current position as research chemist in the Fire Ant Research unit of the USDA. His current research interests center around the complex chemical ecology of fire ants. He is the co-editor of Fire Ants and Leaf-Cutting Ants: Biology and Management (Westview Press 1986). [Insects Affecting Man and Animals Research Laboratory, Agricultural Research Service, USDA, Gainesville, Florida, 32604].

P. KIRK VISSCHER received his Ph.D. in entomology from Cornell University. His research interests span the behavioral ecology of honey bees, from the nepotistic intrigues of kin discrimination to broad cooperative patterns of colonial foraging. [Department of Entomology, Cornell University, Ithaca, New York, 14853].

KEITH D. WADDINGTON is associate professor of biology and a behavioral ecologist at the University of Miami. He received the Ph.D. in 1977 from the University of Kansas. He is interested in the evolution of complex behavioral patterns, including foraging and communication, and in mechanisms which underlie complex behavior. His current research is focused on the

energetics of foraging in bees. [Department of
Biology, University of Miami, Coral Gables,
Florida, 33124].

Index